D0518321

Animal Navigation

ANIMAL NAVIGATION

Talbot H. Waterman

SCIENTIFIC
AMERICAN
LIBRARY

A Division of HPHLP
New York

Frontispiece photo:
Baby green turtle entering the surf.

Library of Congress Cataloging-in-Publication Data

Waterman, Talbot H. (Talbot Howe), 1914-
Animal navigation/Talbot H. Waterman.
p. cm.
Bibliography: p.
Includes index.
ISBN 0-7167-5024-4
1. Animal navigation. I. Title.
QL782.W37 1989
591.1—dc19 88-15811
CIP

Printed in the United States of America.

Scientific American Library
A Division of HPHLP
New York

Distributed by W.H. Freeman and Company.
41 Madison Avenue, New York, New York 10010 and
20 Beaumont Street, Oxford OX1 2NQ, England

1 2 3 4 5 6 7 8 9 0 KP 7 6 5 4 3 2 1 0 8 9

This book is number 26 of a series.

Contents

To J. G.

Preface

Because the roots of this book reach far back into the beginning of my scientific career, *Animal Navigation* reflects a major part of my academic and personal life. Visual orientation in a water mite was the topic of my senior honors research in college and was the subject of my first published paper. John H. Welsh, with whom I did my Ph. D. thesis, and Henry B. Bigelow, the first Director of the Woods Hole Oceanographic Institution, who was a no-nonsense inspiration in marine science, were germinal influences during my undergraduate and graduate years at Harvard. My scientific outlook was also significantly shaped by the psychophysicist S. S. Stevens, despite some resistance on my part at the time I worked with him at the Psychoacoustic Laboratory.

Further World War II operational research on radar at the Radiation Laboratory at MIT and in the Pacific reinforced and broadened my interest in biological navigation. Shortly thereafter as a junior faculty member at Yale, I was profoundly affected by a lecture given by Karl von Frisch in which he described his recent discovery that honey bees use the polarization of sky light in their orientation. Not long afterwards I wrote a review titled "Flight instruments in insects," inspired in part by my wartime concern with aircraft navigation and in part by the excitement I felt about von Frisch's research. Over the years the give and take of my university teaching, particularly in a perennial graduate course on animal orientation and navigation, also fostered and molded my interest. Even now, as Professor Emeritus, I conduct research on the orientation of marine animals to underwater polarized light, trying to discover if they can use it as bees and ants do for spatial orientation and direction finding. I am deeply indebted to many people and many pivotal experiences. Teachers, students, colleagues, associates, and friends have all contributed numerous ideas and queries for which I am most thankful. The various institutions and agencies that over the decades supported the underlying research and study were crucial from undergraduate days on. The Society of Fellows at Harvard, the Office of Naval Research, the National Geographic Society, the Woods Hole Oceanographic Institution, the National Science Foundation, and Yale and Harvard Universities should be specially mentioned.

Those who have collaborated most directly in the preparation of this book and in the procrustean task of editing it should be thanked particularly—Mabs Campbell, my long-time research and editorial assistant; Dan Maffia, for four fine paintings; and not least, several of the publisher's staff: Janet Wagner, for bountiful and often enhancing editorial advice; Phil McCaffrey, for assembling text and illustrations of high quality whose parts all know one another; Travis Amos, for compiling superb photographs; Nancy Field, for book design; Mike Suh and Anna Yip, for overseeing the art work; and Susan Stetzer, for coping with a whirlwind production schedule.

Talbot H. Waterman
New Haven

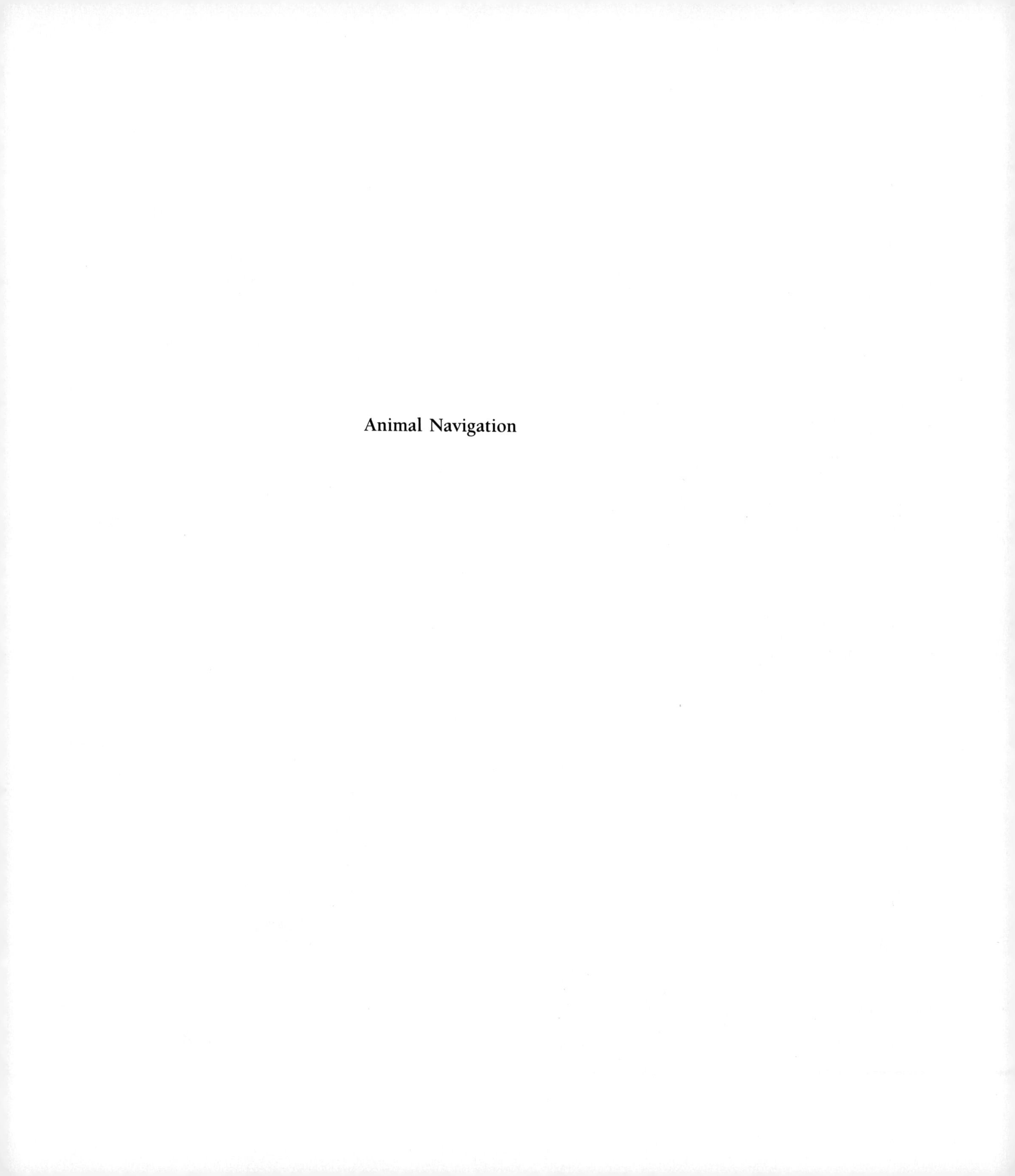

Animal Navigation

Introduction

Our everyday experience is full of animals and people on the move.

Mornings around sunrise sea gulls, alone or in small groups, fly unhurried past my windows. As I start my warm-up exercises and meditation, they are on their way from nighttime roosting areas along the Connecticut shore to daytime foraging and feeding grounds. While they fly 10 or more kilometers inland, past Konald's Pond, Lake Dawson, and beyond, I wonder idly how they know their way. Can they already see their destination when they fly up from their roost? Are they following West River to its source in Bethany Lake? Do they take a fix on West Rock from the harbor and then follow its steep scarp further inland? Or do they have a mysterious "sixth sense," some innate map and compass, which guides them? Have they already explored the area and learned the way? Clearly my early rising neighbors are commuting downtown at the same time by driving along a route made familiar through repeated use.

From my sixth-floor laboratory in Kline Biology Tower at Yale, I used to look down on a big hollow oak that is home and jungle gym for a troop of gray squirrels. This particular tree, one of many growing on what were once the broad sloping lawns of the Hillhouse mansion, has long been the hub of a local squirrel population. From its nest holes baby squirrels emerge each summer, learning to climb and run over its trunk and branches. Later they jump to neighboring trees, then venture farther and farther away from home until they develop their adult territories. But many of them return to this oak for shelter, for fellowship, and seasonally for breeding.

On warm summer days, the sunny window ledge of my laboratory usually has a bee-fly (a near perfect mimic for a bumblebee) poised on its outer edge. Scanning the bright southern sky for some time, the insect suddenly flies furiously into space.

Red-winged blackbirds *Agelaius phoeniceus*, like many bird migrators, gather in ever-increasing flocks before finally departing on their fall mass movement southward.

After a few seconds out of sight, it returns as swiftly to resume its panoramic watch. These forays continue at frequent intervals through much of the midday. According to colleagues knowledgeable about insects, these are mating flights in which the male fly waits on the ready until he spots a passing female of the species. Then he pursues, intercepts, and, if the foray is successful, copulates. My observation that the males returned with abdomens pumping, a little breathless from their flight, yet never brought back traces of captured prey certainly fits such an explanation. But only rarely was I able even to see that the fly was pursuing something, let alone be sure that its target was indeed an attractive female bee-fly!

It is tempting to tell, too, of the thousands of starlings that roosted nightly last fall in those same Hillhouse oaks. Each morning they dispersed widely to feed, returning again soon after sunset to spend the night in a dense jittery flock. Then they were gone for the winter. I could also recount the mile-long straggling flight of red-winged blackbirds that passed daily over the Tower at sunset for most of a week in November, heading easterly toward a roost apparently near East Rock or Sleeping Giant Mountain. No doubt they were either joining a huge flock mobilizing to head south or just waiting for the weather to urge them on to the rest of their long journey. Such anecdotes remind us that most animals (including humans) spend much of their time in transit—whether swimming, flying, walking, creeping, burrowing, jogging, running, galloping or jetting, from one place to another. All this movement depends mostly on the interaction of two factors. First, the earth is a mosaic of localized sharply different places: mountaintops, meadows, hot springs, glaciers, beaches, forests, deserts, deep seas and so on. Second, although many animals may prefer one of these habitats, they must move through the larger environmental patchwork to meet their complex needs. The mountain goat, for instance, may thrive in high alpine meadows in the summer but has to come down to lower levels in winter. Deer that browse in the fields at night hide in the forest during the day. The best place for an animal to feed is usually different from the best place for it to sleep, and the best place to build a nest is ordinarily still somewhere else. Furthermore, the suitability or availability of these special regions often changes regularly with the seasons, the time of day, the tides, and even the organism's age. Accordingly, a mobile animal's habitat typically consists of a number of subhabitats. Each of these has certain features that satisfy particular biological needs at a specific time or during a certain phase of the life cycle.

To move between subhabitats efficiently and safely animals must travel in the right direction, to the right distance, at the right time. How they are regularly able to reach these objectives is the subject of this book. *Navigation* enables animals to find their way from one place to another. It is a regulated, nonrandom activity that improves their chances for survival in the earth's spatial and temporal mosaic. In this book animal "objectives" and "goals" do not imply self-awareness or intentional planning on the organism's part. Rather the objectives and goals are considered to be the "set points" of the organism's programmed behavior much as the temperature

Because loons have weak lifting power for their flight and legs and feet that are set so far back on their bodies that the birds can scarcely walk, they cannot take off from land and need long expanses of water as well as high air speeds, perhaps 35 kilometers per hour, to become airborne. When the remote wilderness lakes in the northern United States and Canada—the usual summer habitat for the common loon *Gavia immer* in North America—are frozen over, these fish-eating divers must migrate seasonally to seek open water as well as available food.

dialed on a thermostat normally brings the system it controls to the heat level indicated. Obviously, animal navigation is far more complex than a simple temperature regulator. Yet the analogy serves because our main purpose is to determine how the system works rather than why or how a certain objective was set for it. For instance, the means whereby an arctic tern can navigate halfway around the world from the antarctic pack ice, say, to the coast of Greenland are surely a different matter from why it happens to breed in Greenland and to live during the nonbreeding season in Antarctica. For animals as for human pilots, finding the way is quite distinct from the urge or decision to move somewhere else. Animal navigation will accordingly be explained here mainly in terms of *how* it is done with only occasional attention, as in the last chapter, to *why*.

When distances are short and the animal can, for instance, see or hear its goal, a formal navigation program may in human terms seem trivial. Yet at night, in fog or whenever the goal is far away, navigation is surely needed. At certain times finding the way requires skill that seems to verge on clairvoyance. However, sophisticated *human navigators* ordinarily take a matter-of-fact approach to their job. Because we know exactly how human pilots navigate, reviewing their methods should be useful in trying to understand how animals reach their destinations. Human navigation is made up of four mutually supportive procedures. One or another of these may suffice under favorable circumstances. However, in a pinch, the navigator exploits every available means to reach his destination.

The first procedure is *piloting,* in which known natural landmarks or seamarks (like a submerged reef) as well as markers of human origin such as buoys, lighthouses, and airway beacons are used sequentially in coastal voyaging or overland flight in clear weather. Shorelines, rivers, and mountain ridges clearly serve certain animal navigators in the same way. Hawks and eagles migrating along mountain ridges and the gray whales' close following of the North American Pacific coastline are examples. Provided that visibility is good and familiar landmarks plentiful, human piloting in well-known territory is simple and readily learned. Charts and maps are usually important adjuncts to memory. They are essential in places not known to the pilot. Particularly on land, where there are well-posted roads or trails, we are all quite used to piloting when driving or hiking. Difficulties obviously arise when landmarks are either forgotten or hard to identify and signs are illegible or missing. If such clues are absent or unknown, piloting must be replaced by exploration. Then common sense as well as rules of thumb, like always head downhill or turn right at any fork, apply. Ordinarily after such path breaking, the ability to return home is critical.

The second navigational procedure, *dead reckoning,* can be used without landmarks—on or under the open sea, in fog, in the air above the clouds—and in exploring the unknown. To use dead reckoning the navigator measures the direction

Migratory routes often follow coastlines like the one in this air view of the Peninsula de Quevedo on the Pacific coast of Mexico, just north of the Tropic of Cancer.

An example of dead reckoning. Suppose that a ship sails due west from San Francisco and the Golden Gate *A* for an estimated distance of 100 kilometers; then at *B* it changes course to southwest for another 100 kilometers; finally at *C* it again heads directly west for still another 100 kilometers to *D*. The navigator can readily calculate his ship's position at *D* as 280 kilometers from *A* in a direction nearly west by south. If the latitude and longitude of *A* are known, the location of *D* is easily determined. Usually, however, winds and currents cause errors, which must be corrected with other kinds of information.

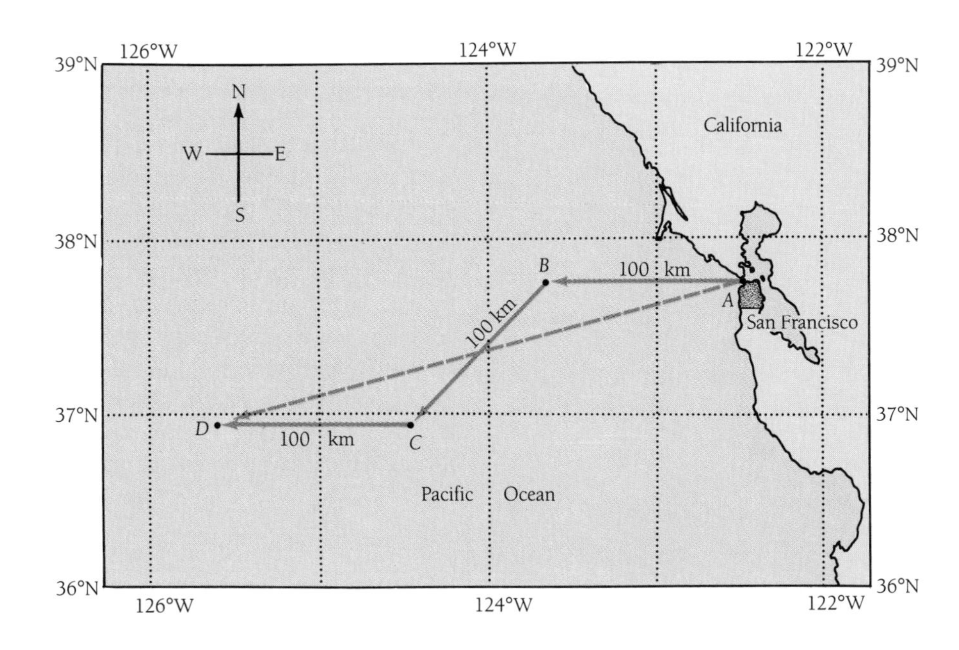

Classic tools of a ship's navigator include charts, a compass, a chronometer, a sextant (center) sailing directions, astronomical tables, and various measuring and drafting instruments.

of travel and the distance covered on each leg of the course. While advancing, he repeatedly adds the measurements for a given segment of the journey to the preceding one to yield the current position relative to the starting point. This process is repeated until the destination is reached. Note that a compass and a clock, as well as ways of measuring speed and drift, are essential to the human pilot. Although they lack real instruments and graphic maps, animals are known to use dead reckoning often.

The third procedure is *celestial navigation* in which the sun, moon, planets, stars, or sky are used to obtain positional, directional, or temporal information. Locating north from the bearing of the polestar, fixing one's location by calculating its latitude and longitude from the sun or stars, and determining local noon from the moment the sun reaches its highest point in the midday sky are common human uses of celestial observations. An accurate clock, celestial charts and tables, as well as such instruments as a sextant, are typically needed for this kind of navigation. The sun's direction and other data derived from the sky help orient and time the movements of the honey bee and migratory birds as well as many other kinds of animals.

Finally the fourth human procedure is *electronic navigation*. This is widely used in direction finding, obstacle avoidance, runway location, latitude and longitude determination, and timekeeping. The great advantage of such systems is their independence of time of day and weather. Nowadays sophisticated methods, such as finding one's location by using computer-processed satellite signals, tend to overshadow the first three, much older, procedures even in international yacht races.

However, no animals can conceivably have even the simplest instruments required. As a result high-tech navigation of this sort is interesting to us now mainly by analogy—between, for instance, a ship's sonar and a porpoise's echolocation.

On the other hand, animal migrators clearly do use additional navigational aids not typically exploited by us. Some of these depend on certain senses, like olfaction, critical for bloodhounds but not for jet navigation. Others take advantage of several geophysical factors that animals can sense. For example, the earth's magnetic field, which by causing the compass needle to indicate north, has played a key role in human navigation since the late Middle Ages, has in more recent times often been considered as a possible source of information for animal path finding. Other obvious and often important directional signals may come from winds or currents. Swells on the open ocean, which often maintain a steady direction over large areas and times, may serve to orient long-range migrating fish and birds. (In traditional human canoe navigation wave patterns do indeed provide important data.) Other

As air-breathing mammals, porpoises are remarkably successful aquatic animals, fast swimming and wide ranging in the world's oceans. Here near Hawaii a group of spotted dolphins *Stenella attenuata* dive in pursuit of prey (fish or squid). Porpoises and dolphins skillfully use echolocation for short-range navigation, but their long-range methods are largely unknown.

less familiar geophysical clues for animals may include the Coriolis force due to the earth's rotation. This acts to deflect moving objects including animals—clockwise in the northern hemisphere and counterclockwise in the southern. Its strength decreases as one moves toward the poles. The earth's gravitational field, sustained electrical potentials in water (some due to fluid flow), local anomalies in the earth's magnetic field, the atmosphere's barometric pressure, and infrasound generated by storms or surf (of very low frequency and inaudible to us) may all assist animal navigation. Infrasound carries easily over long distances in the atmosphere and is readily detected by some birds, at least. The intriguing possibility that animals use these geophysical clues to navigate remains in many cases incompletely documented. Both the senses involved and the evidence for their application need intensive research. Undoubtedly some of the mysteries and controversies of animal migration could be dispelled by first identifying and then confirming the importance of these and still other subtle information sources that may remain to be discovered.

Having asked some pertinent questions, we may now recall our most basic definitions. How animals find their way from one place to another is what we mean by *animal navigation*. We should add to this "efficiently and safely" because their navigation is an adaptive process that increases a migrating organism's chances for survival and reproduction. *Migration* as used in this book means periodic cyclic movement from one part of an animal's habitat to another and back. Animals may migrate between two remote summer and winter living areas, like Alaska and Central America, or among a series of subhabitats. There are many patterns that we shall encounter. Because animals migrate, we know that they navigate. Solved navigational problems are evident in the routes and timing of migrations made by many insects, fishes, birds, and whales. The duration of the round trip, the use of one or more way stations, and the distance traveled all vary widely from species to species as does the biological reason for the journey. Typically though, migration is a seasonal, annual phenomenon closely related to reproduction, development, feeding, climate, and weather. But some species may complete only a single migratory cycle in a generation lasting a number of years; examples are the Pacific salmon and American eels, which die after spawning once. In contrast, some insects require more than one generation to complete a single seasonal circuit.

Usually migration implies rather long distances traveled, but what "long" means may be quite relative. For instance, a very long goal-oriented movement for a microscopic *Paramecium* would not have to reach more than one or two body lengths of a blue whale! Obviously the actual size of animals strongly affects their migratory range. Speed and efficiency of locomotion typically increase rapidly with size. The swimming speed a trout can sustain, for instance, varies with its length, multiplied by three or four. Thus a fish that is twice as long as another can swim six to eight times faster. Running and, to a lesser extent, flying speeds are also typically scaled (within limits) to animal size. Larger animals are consequently more likely to migrate globally than smaller ones are. The cost of locomotion—that is, the energy required

A column of African driver ants *Dorylus sp.* on the march; workers are flanked by ranks of quiescent nest mates on either side. Such colonies form fearsome raiding parties that range out from their nests by the millions in thick black strands, killing all the insects and even some lizards or snakes they encounter. Some ants use sun and sky for direction finding, but driver workers are eyeless and must depend, like many other ants, on contact with their fellows and chemical markers for trailblazing.

per gram per kilometer—decreases steadily with increasing size for animals just as it does for ships and aircraft. Hence larger animals can move about more efficiently (as well as faster) than smaller animals with the same kind of locomotory system. Energy costs vary among different types of locomotion as they do among sizes. Flying is significantly faster but generally more costly than swimming because, among other things, work must be done continuously to support an animal's weight in the air. Even so, flight is substantially cheaper energetically than are walking, running, and other forms of terrestrial locomotion because swinging the legs back and forth uses a lot of energy.

Because of energy costs, global migrators are typically swimmers and flyers that are not too small. For the same reason, terrestrial animals do not usually travel over such spectacular distances as birds, fish, and whales do. Nevertheless the terrestrial migrators with the longest ranges are large animals like reindeer, caribou, or antelope. These may walk, rather slowly, hundreds of kilometers between seasonal ranges. Despite the general migratory limitations of small size or less than persuasively athletic locomotion, we should not overlook the long-range prowess of the Chinese wool-handed crab, the monarch butterfly, the eel larva, and the ruby-throated hummingbird. Even with their apparent handicaps, all are remarkable migrators on a world scale, as will be discussed in the first two chapters. An important further interplay between size and locomotion relates to the influence of winds,

A column of migrating American elk *Cervus canadensis* crosses a ridge in the Tetons near Jackson Hole, Wyoming. Movements of such large animals have been recorded in detail by satellite monitoring of individuals wearing collars outfitted with radio transmitters. Piloting by landmarks seemed important for this migration, but the distance covered between winter and summer areas was well under 100 kilometers.

water currents, and turbulence on animal movement. Such small, weakly swimming or flying types as eel larvae or midges are largely at the mercy of commonly present flows of water and air. Accordingly, these drifting animals may directly make use of currents or winds to carry them where they need to go. Even large, powerful swimmers and flyers have evolved routes that take advantage of the regular flow of their

Milkweed bugs fly north on their spring and summer migration; two or three generations are usually needed to complete the journey because the insect's life expectancy is shorter than the time required. Inherited information seems essential for the insects to know which way to fly.

medium to decrease the cost or to increase the speed of their migrations. In fact, human navigators also adjust the courses of high-flying powerful jets, to minimize head winds and maximize tail winds.

A rather intriguing limitation on long-distance migration for really small animals is their typically short life span. Certain insects, such as the milkweed bug *Oncopeltus fasciatus* in North America, make rather extensive northward migrations during spring and early summer. Yet completing the journey may require two or three generations! This chain ends in the fall when reproduction is suppressed, and a return flight to the south begins.

Animal tracks cover the world with a dense interlacing network whether in the air, in water, or on land. A preliminary explanation of these movements is in order even though the question "why" is the central topic of the book's last chapter. Ultimately the reasons for all this animal traveling are environmental and evolutionary. Migration basically allows animals to live more successfully than they could if they stayed in one place. In a uniform habitat with adequate forage or prey, wider movement increases the browsing or hunting range and provides an animal with more food. In patchy environments, which are more usual, systematically feeding in one highly productive site (like a water hole or a hedgerow of flowers) after another is nearly universal. Where two feeding locations provide good foraging at different times, migration between them can be strongly beneficial if the travel itself is not too costly or dangerous. The vast invasion of the Arctic during its brief summer bloom is a striking example. Then snow- and ice-free areas permit fresh plant shoots, seeds, berries, insects, fish, and other foods to be briefly abundant. On the other hand, appropriate movement allows animals to evade hostile or even lethal conditions like the antarctic winter or the dry season in the Central African plains. Even when there are regular daily or tidal changes in the environment, short-range migration permits the exploitation of various regions at their best. Shore birds, for instance, often feed mainly on exposed mudflats in the period approaching low tide when many worms and clams are freshly available. Insect-eating bats also time their foraging migrations to coincide with the twilight flights of their prey.

If successive stages of an animal's development, such as larva and adult, need different habitats, migration from one to the other is important for the success of the species. Thus fish often have distinct spawning, nursery, juvenile, and adult feeding areas, as well as another for overwintering. The same is true for bats, which often carry out different activities like sleeping and parturition is quite separate locations.

The brief summer abundance of the arctic tundra is witnessed by this rich carpet of moss and other plants in the Yukon on which a pair of black turnstones *Arenaria melanocephala* have made a nest that contains four speckled eggs. Many other bird species migrate to such high latitudes to breed from northern wintering sites as far away as Antarctica, Africa, and the South Pacific.

Most global migrators are swimmers and flyers not only because energy efficiency strongly favors these kinds of locomotion but also because of the continuity of the atmosphere and of the world's oceans. Our account of migration begins with spectacular examples of flyers in Chapter 1 and continues in Chapter 2 with swimmers. The routes and timing of such behavior vividly illustrate the kind of global navigational problems that are routinely overcome by insects, fishes, birds, bats, and whales. Next, to give some insight into how animals might do it, the techniques humans use to navigate without a literal compass, map, or clock are outlined in Chapter 3. Amazingly, native canoe pilots in the central Pacific can still employ ancient methods of interisland voyaging, which rely on keen senses, much traditional, but not written, learning and protracted hands-on experience. Even so, there is no evidence that such unsophisticated humans have ever carried out the rapid, repeated, long-range migrations typical of wheatears and bobolinks or tuna and salmon!

A navigator must be able to sense direction, distance, and time. The central block of chapters considers how animals control their goal-seeking locomotion, particularly with regard to these three requirements. To begin with, Chapter 4 analyzes the importance of spatial orientation (which way is up?) and how to maintain a steady course—functions basic to their navigation. Gravity-sensing systems and visual responses to movement are found to be prominent in this topic. Then Chapters 5 to 7, document the many compasslike elements known in animal behavior. Because all long-range animal navigators have well-developed eyes, a strong emphasis on vision is not surprising. Particulary intriguing among the uses of vision is the good evidence for celestial navigation by many animals including insects and birds. Other senses also provide essential compasslike information, such as chemoreception in salmon homing and moth mating. Electroreceptors and magnetoreceptors fall within a category of geophysical sensors of which we ourselves have no subjective experience. Nevertheless, electric and magnetic direction finding have been clearly demonstrated for several animal groups, at least in the laboratory.

The *map sense,* an awareness of place that goes beyond direction finding, fulfills a basic navigational need to be constantly aware of one's location (Chapter 8). Clearly almost all animals have an internal local map sense because they have a home territory that they know or learn. In addition, some species can return home after moving or having been displaced beyond previously visited areas. The homing pigeon is the best known of such examples and has been a favorite subject for studying animal navigation. Over a short range at least, a number of animals can tell where they are by "reading" some sort of internal "pedometer." This reading is based on information that the animal collects while on the outward path, and the mechanism involved is reasonably understandable to us. However, the really challenging task is to identify one's position by sensory data available only at the site itself.

Following the lead of the geographers of ancient Greece, human map makers and navigators use a two-dimensional grid to specify the position of any point on earth. One of these coordinates, *latitude* is measured in degrees north and south from the equator, which is assigned to be zero. Accordingly, the *tropics* extend from 23.5°N to 23.5°S latitudes, the arctic and antarctic circles are at 66.5°N and 66.5°S, and the poles are 90°N and 90°S. The temperate zones where most of us live lie between latitudes 23.5° and 66.5° in both hemispheres. Perpendicular to the equator are a set of imaginary great circles that run through both poles. These define a second coordinate, the *longitude*. Zero longitude has been arbitrarily chosen as the half-circle passing through the Royal Observatory in Greenwich, England and is called the prime meridian. Other longitudes are measured east and west of Greenwich up to 180°, the International Date Line. By common consent the world's day ends just east of that line in the mid-Pacific Ocean and the next day begins just west of it. Global travelers lose a day crossing the dateline between Honolulu and Manila but gain one on the return trip. Whether living things (other than humans) have an abstract map sense in which such coordinates identify places they have never experienced before, remains to be proved. Hence a major question about animals as navigators is: Can they possibly have some comparable two-dimensional reference system, perhaps only some sort of crude geographic coordinates?

Chapter 9 discusses the means by which animals are provided with the clocks and calendars needed for their navigation. Nocturnal migrators like songbirds, for example, must start their nightly travels at the right time. Similarly, the shad must begin to swim upstream at the right season for successful spawning. Finally Chapter 10 turns away from a preoccupation with *how* animals find their way to conclude with a brief discussion of *why* they move about so much and why they have particular destinations.

1

Animal Migration: Flyers

Air and *flight* seem the epitome of freedom. Aptly, flying animals are among the most liberated of travelers; they vigorously seek food, breeding grounds, and living space around the world and from pole to pole. Their migrations display their skills as navigators. At present three quite distinct kinds of animals on earth can sustain free flight. Among birds, bats and insects there are accomplished aerial navigators, some of which essay extraordinary migrations.

BIRDS

The popular impression that certain birds are the nonpareil of migrators could be supported on several counts. Because they fly easily above both land and water, birds range freely over the world. Let's begin with some world class navigators. The gold medal for migration could be awarded for any of several particular virtues such as speed, efficiency, accuracy, safety, or long-term survival value to the species. If we take *distance* as our criterion, the clear winner is the arctic tern *Sterna paradisaea*. Its habitat covers arctic and antarctic living zones connected by three or four migratory flyways. Two destinations could not be farther apart unless the birds left the earth. After breeding during the northern summer, sometimes within a few degrees of the north pole, this doughty navigator then flies all the way to the Antarctic. There the birds overwinter in the southern summer near the edge of the pack ice. Upwelling of deep water around the south polar continent makes the area a particularly rich feeding place. When they return to the high Arctic to breed again during the next northern summer, the birds complete a round trip of 30,000 to 40,000 kilometers. Because of its bipolar, double-summer habitat, the arctic tern spends about 8 months a year in continuous daylight, no doubt another record.

Powerful and graceful flyers, adult arctic terns fly round-trip each year between their arctic breeding grounds and marine wintering area near the antarctic pack ice, nearly spanning the globe north and south.

The actual path followed in this interhemispheric journey depends on where individual terns breed. Birds that have nested at high latitudes over eastern North America cross the Atlantic to join terns that nested in northwestern Europe. Together they fly south to reach the bulge of West Africa where they split into three groups near Dakar and Cape Verde, just before crossing the equator. One group continues to follow the African coastline, another heads directly into the open South Atlantic toward Antarctica and the third flies southwest perhaps 2500 kilometers across the equatorial Atlantic to follow the South American coast to Cape Horn. The smaller numbers of arctic terns that nest in western North America and eastern Asia begin their southward migration by converging on southern Alaska. From there they skirt the western coastlines of North and South America, flying southeastward to the lower tip of South America. In the final stressful stages of their journeys all four groups head out over the sea south toward the still distant pack ice around Antarctica. Once in high southern latitudes they continue to fly eastward to congregate in the main wintering area south of the Indian Ocean. Some individuals, particularly young birds, continue a downwind circuit around the world helped by the strong prevailing westerly winds. In the spring arctic terns from the main Atlantic group and those from the smaller Pacific group return to their northern hemisphere breeding grounds by different routes, the actual paths of which are still not well known.

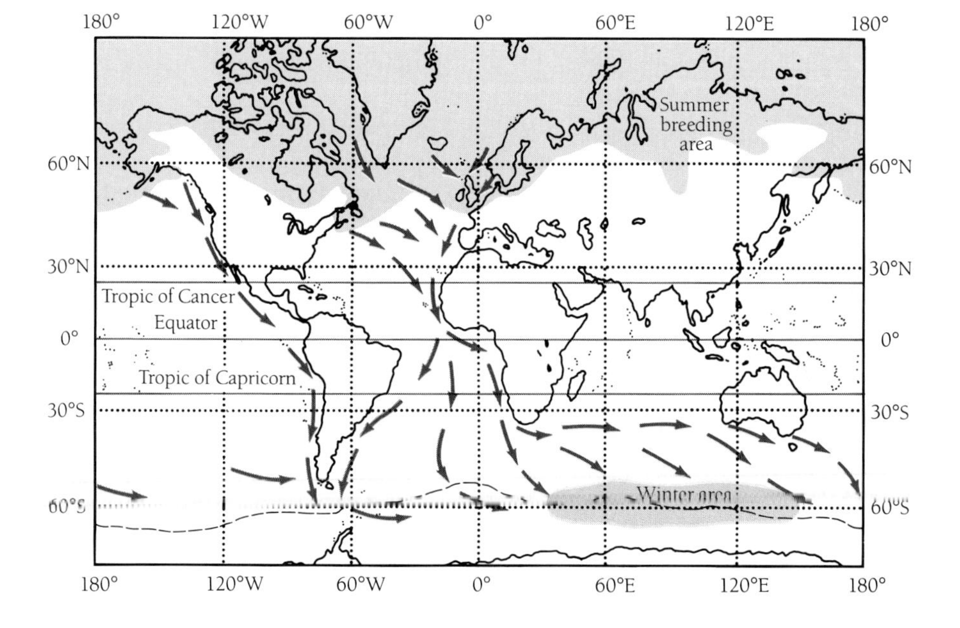

Breeding areas of the arctic tern (green shading) are connected by several southward flyways (arrows) to the wintering grounds (blue shading) around Antarctica, mainly south of the Indian Ocean, where they feed on an abundance of plankton. Some individuals, mainly juveniles, also drift and fly eastward around the world at these high latitudes. The return paths followed at the end of the antarctic summer to reach the northern spring again are not well known.

Other long migrations

Although the arctic tern holds the record for distance traveled, quite a few other migratory birds undertake spectacular if somewhat shorter round trips. For instance, the lesser golden plover *Pluvialis dominica* commutes across the equator between widely separated subhabitats each year. Different individuals of this species migrate overland to southeast Asia, Indonesia, and Australia from high-latitude summer ranges on arctic and subarctic tundra extending from Asian Siberia to Alaska. Others from eastern Siberia travel south over the Pacific as far as New Zealand and Tasmania. Some other lesser golden plovers fly from western Alaska directly out to sea more than 3000 kilometers to Hawaii. From there they continue over a large area of the South Pacific. Birds from more central and eastern North America nesting grounds travel 2000 to 3000 kilometers south over the North Atlantic, departing

(*Left*) A Pacific golden plover traveling from Alaska or the western Arctic stretches and preens on French Frigate Shoals, Hawaii before continuing further south into the Pacific or to southeast Asia.

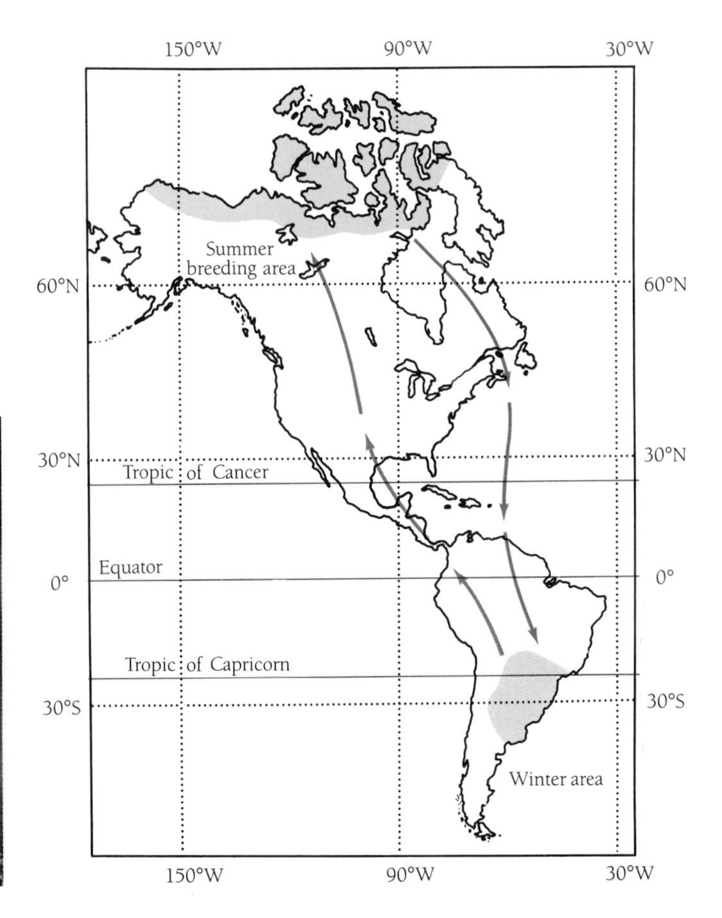

(*Right*) Lesser golden plovers that breed in the eastern Arctic migrate southward in the fall over eastern Canada and out over the open ocean nonstop to the coast of South America continuing far south to their winter area. Like many migratory birds their continental return route is a large loop. Differences between fall and spring wind patterns, climates, and food supplies probably account for such distinct southbound and northbound pathways.

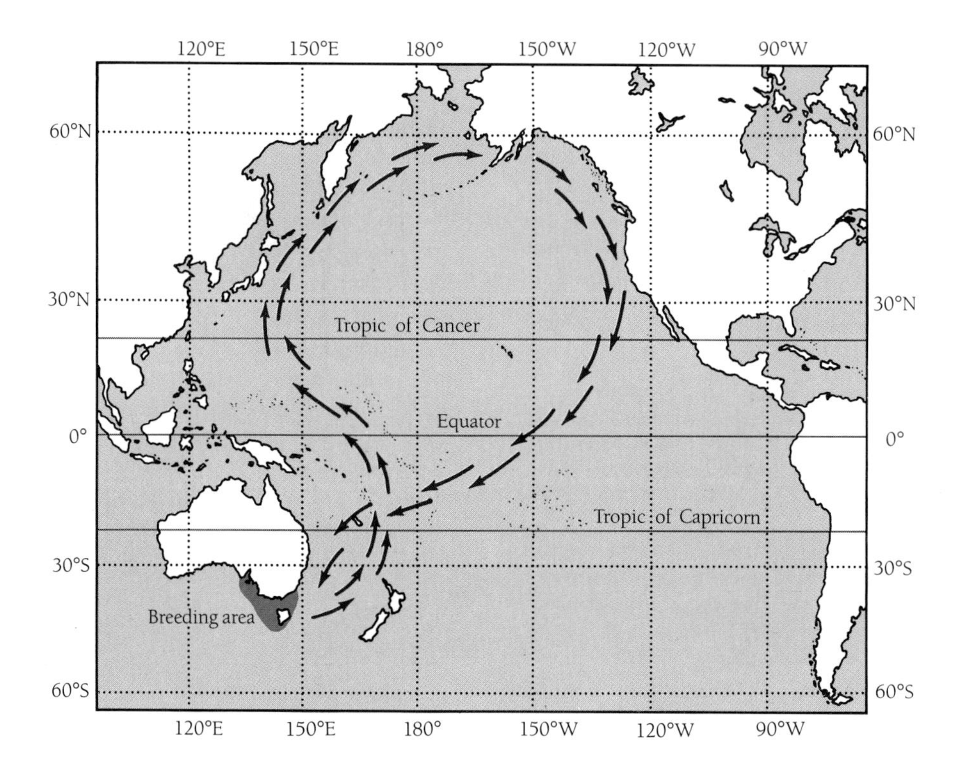

The short-tailed shearwater migrates across the Pacific from its breeding grounds off the southeast coast of Australia. Flying far north along the coast of Asia, many individuals of this species travel as far as the Arctic Ocean. Their ongoing course skirts western North America and recrosses the Equator in the central Pacific. Reaching the Australian east coast in Queensland, it follows the shore southward to the nesting area. The full circuit covers 20,000 to 35,000 kilometers. Like the routes of many far-ranging migrators, this shearwater's long feeding track exploits prevailing wind patterns.

from Canada's Maritime Provinces. This extended overseas flight reaches the equatorial north coast of Brazil near the mouth of the Amazon, and continues to wintering grounds in southern Brazil and Argentina. In the spring these lesser golden plovers complete a long migratory loop by flying north over South America, Central America, Mexico, and the Mississippi Valley en route to the Arctic.

Some far-ranging oceanic birds have migration cycles with "reversed" polarity. Instead of breeding in the north and wintering in the south, they breed in the southern hemisphere and fly long distances to high northern latitudes to feed at sea and mature. Quite spectacular, for instance, is the cycle of the short-tailed shearwater *Puffinus tenuirostris,* which nests early in the austral summer. Birds native to Bass Straits and adjacent Australian shores set out on a migratory circuit around much of the Pacific. Over a period of several years they follow a lopsided figure-eight course that includes Japan, the Arctic Ocean, the California coast, and the South Pacific before finally returning to their nesting grounds. The migration of the great shearwater *Puffinus gravis* in the narrower Atlantic basin is somewhat similar. From breeding grounds on the south Atlantic islands, Tristan da Cunha and, in smaller numbers, the Falklands, it flies a large figure-eight pattern reaching far into the North Atlantic during the northern summer. A close relative, the main subspecies of

Abundant food in the Peru Current off the west coast of South America has made it a profitable migratory destination for many oceanic birds in their nonbreeding seasons. Species from the Arctic to the Antarctic and from North America to the far southwest Pacific converge on this area by navigating along the five different tracks shown here, which outline the routes of sixteen different species.

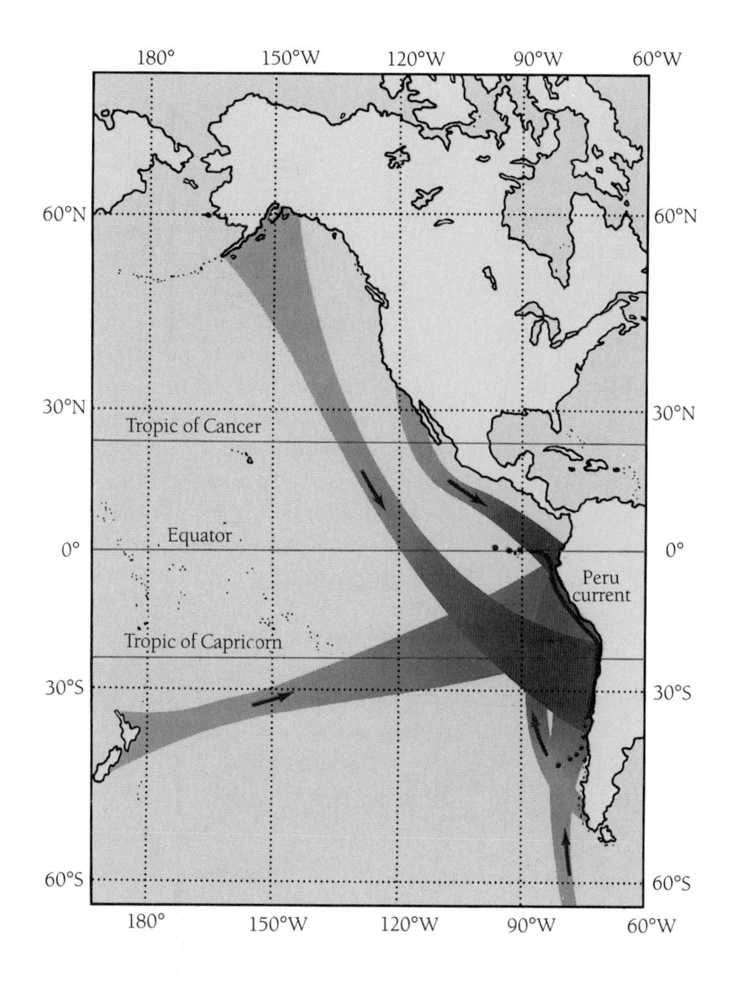

the Manx shearwater *Puffinus puffinus,* inverts the hemispheric allegiance by breeding in the eastern North Atlantic (British Isles, Iceland, the Azores, and Madeira) and wintering in the western South Atlantic.

Seasonal availability of food and nesting sites has shaped the migration patterns of many oceanic birds. Long-range flyers converge on certain prime breeding or feeding places at appropriate times of the year. One of the most productive oceanic areas in the world is off the west coast of South America, for example, where the Peru Current flows northward. The high productivity is due to upwelling of deep water laden with inorganic nutrients necessary for the growth of algae, which form the first link in the animal food chain. The resulting bloom supports a burst of rich animal plankton that in turn provides food for important fisheries as well as enormous populations of foraging seabirds. Among these, skuas, petrels, gulls, and shore birds come in seasonally to feed, migrating from their particular nesting grounds in the Arctic; other species come from the California and Baja California coasts. Franklin's gull *Larus pipixcan* flies in during the northern winter from its breeding areas in the North American great plains. Others migrating from Antarctica and the southern tip of South America include a shearwater, two species of petrels, the great skua, the cape pigeon, and prions. From New Zealand and its outlying islands two albatross species and a petrel also feed in the Peru Current. Clearly the timing of all these convergent flights depends on the birds' various seasonal breeding periods.

The availability of nesting grounds, like that of feeding grounds, may also determine the migrations of several other Pacific marine birds. For instance, five different species (a gull, two petrels, a booby, and an albatross) all breed in particular island sites within the Galapagos Archipelago. Like the great majority of sea birds, they nest in dense colonies. But each species has its own separate, remote feeding area. Hence the compass directions, timing, and distances in their round-trip migrations are quite different for the five species despite their close nesting sites. Closely related species of oceanic birds may differ not only in their pattern of migration but also in whether they migrate at all. For example, within the genus *Sterna*, the arctic tern, as we have seen, is a champion world traveler. Yet the Kerguelen tern *Sterna virgota* lives closely around a few isolated islands far south in the Indian Ocean where it breeds. Other species of *Sterna* show a full scale of intermediate migration ranges. Similarly, some shearwaters in the southern hemisphere differ sharply from their ocean-spanning relatives by always staying close to home.

Even within a given bird species, juveniles and adults may have different migratory patterns. Young terns and short-tailed shearwaters typically linger in their wintering range for an extra year or two before returning for the first time to the breeding area. The first-year young of the European swift *Apus apus* and the white wagtail *Motocalla alba* start to migrate in the fall before older, more experienced birds do. In contrast the young of the Hudsonian godwit *Limosa haemastica* and the golden plover remain in the breeding area until after their parents have gone. Banding recoveries of European white storks *Ciconia ciconia* have proved that even nest mates may follow remarkably different migration routes. One route of white storks leads to Africa via Iberia and the Straits of Gibraltar while another passes over the Bosporus and the Near East. These two detours from the direct route to Africa from the same starting place bypass the Mediterranean Sea. The source of this duality is puzzling. Does each bird have both flight plans? Or do some have only one?

Storks routinely use rising air currents to make long flights during the day. Like many other large birds, cranes and storks depend strongly on soaring on thermal updrafts for their migration. They can soar overland only during the day in a warm climate. If they cannot see land on the other side of a body of water, they appear loath to fly out over water, where the rising warm air masses required to soar are absent. In contrast, albatrosses and other oceanic birds that use quite different ways of soaring to sustain flight over water spend most of their life at sea in rough windy

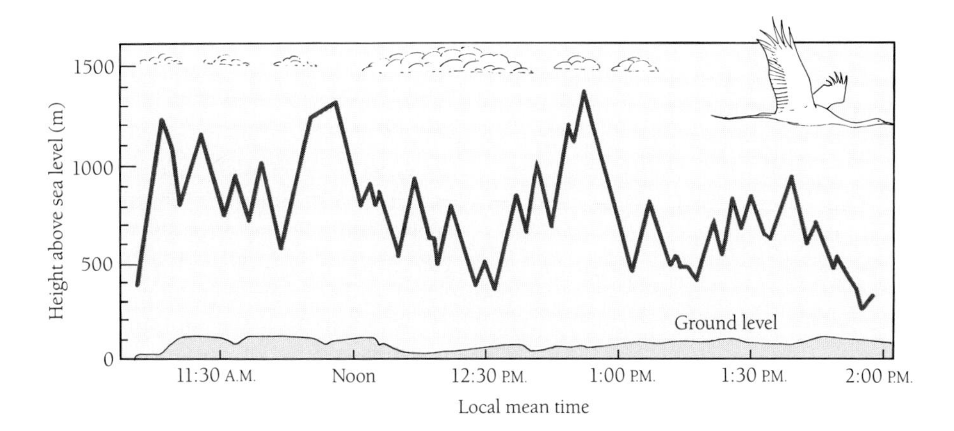

The flight profile of a flock of migrating cranes *Grus grus*. In warm climates the sun's raising the land temperature in clear calm weather generates the thermal updrafts exploited by large migrating birds. Because clouds and turbulence interfere with the formation and persistence of suitable thermals, soaring birds lose altitude where there are no updrafts. In this profile, sharp drops occurred just before and after noon and just after 1 P.M. Near the end of the recorded period the cranes were lower than 200 meters above sea level and *for the first time* since taking off had to flap their wings. Birds also soar on updrafts formed where the wind blows against fixed objects such as cliffs, ridges, and even buildings. Such free lift along mountain ridges is critical for the long-range migrations of many hawks and eagles.

areas far from land. Such birds are rare above the calm windless seas of the doldrums, just outside the tropics.

Most small land birds cannot soar at all. Yet surprisingly, they show little hesitation in crossing substantial "nonstop" areas such as the Mediterranean or the Sahara. Even the ruby-throated hummingbird, despite its tiny size and high rate of metabolism, often migrates 800 kilometers or more straight across the Gulf of Mexico. Many other birds fly directly across such mountain ranges as the Pyrenees; indeed the towering Himalayas are overflown by ducks and geese along with a number of small songbirds, all migrating back and forth from central Asia to India and elsewhere.

The wandering albatross *Diomedea exulans* travels a west-to-east route similar to the most southerly part of the arctic tern's southbound migration. With a wing span of 3.6 meters, its skill at gliding allows it to soar above the seas between 40 and 50° latitude in the southern hemisphere, an environment characterized by strong winds and high swells.

Staging

When the terrain allows it, land birds often migrate in stages. Their overall flight may be made up of a series of relatively rapid, discrete steps punctuated by rest stops. These often are needed for feeding to restore the resources necessary to travel farther. In late winter and late summer many migrators substantially increase their available potential energy by storing fat. These fat reserves may reach up to 50 percent of a bird's body weight before migration begins. Estimates of the maximum distance a bird can fly without further feeding can be made from the amount and specific chemical nature of this stored fuel as well as the metabolic cost of flying at migratory speeds. Obviously foraging en route is necessary if the track to be flown is longer than the reserves can support. The plankton-feeding red-necked phalarope *Phalaropus lobatus* migrates 6000 kilometers from the Canadian subarctic tundra to the open Pacific off Peru. In the course of this flight it must stop on the way to forage at least twice, including an initial period of 2 or 3 weeks, to build up the necessary energy stores.

Many species of birds depend on marshlands and inland waterways along migratory flyways for safe resting and feeding en route; sometimes the migrators arrive in enormous numbers as did these snow geese, stopping over at Tule Lake, California.

Most small song birds stop to feed and rest each day during their migration. Ordinarily their migratory flights occur only at night. If crossing a desert or body of water, like the Sahara or Mediterranean Sea, takes more than half a day, these birds must fly nonstop for 24 hours or more, or until fatigue sets in or their reserves fail. Other birds such as swallows, martins, as well as soarers such as storks, pelicans, and hawks normally migrate only during daylight. Waterfowl on migration predictably pause at certain prairie or coastal marshes. Shore birds, too, commonly proceed from one favorable staging place to another on their journey. Bird banders, as well as hunters, have long taken advantage of such known stopovers.

First-year migrators, often easily recognizable by juvenile plumage, frequently travel without experienced adults. Because the young of certain species leave before or after their parents, their navigation cannot be guided by parents. Some evidence shows that birds that have flown a course before can navigate more direct flights. Veterans tend to get there first.

Accuracy

Departing at the right time and arriving at the right place at the right time are critical tests of animal navigation. Substantial misadventures do, in fact, occur during migration, especially in bad weather. On foggy nights thousands of birds may flutter around a ship at sea or dash themselves against lighthouses, high bridges, or skyscrapers, confused, no doubt, by their bright lights. Violent storms and strong headwinds can cause even greater losses. Birdwatchers may be thrilled to spot a single errant bird hundreds or even thousands of kilometers from its species' normal range. Such strays may never get back on their normal navigational course. For example, a black-browed albatross *Diomedea melanophris* was repeatedly sighted over a period of about 30 years in the Faroe Islands (between Scotland and Iceland). This species is normally limited to cold oceanic areas of the southern hemisphere. If such accidental wandering leads a pair or a population into a favorable new habitat, colonization may occur. Such immigration is well known within historic time as in the invasion of the eastern North Atlantic by gannets or the spread all over the United States of the European starling from a few birds deliberately released in New York City.

Studies testify to the extraordinary precision of many long-range bird migrations. Beyond casual observation that an apparently familiar pair of robins or wrens come back every year to the same shrub or birdhouse in our garden, there is much detailed evidence. Individually banded birds have homed to highly specific locations for many years in a row. Some seabirds returning to a particular rookery on a certain small island may be so precise that a single persistent burrow is repeatedly used for nesting by the same pair after return flights from wintering ranges thousands of kilometers away. Some hummingbirds retain an amazingly detailed memory for particular feeding areas to which they come back each spring. Such site fidelity to local feeding territories has also been observed for some species in the nonbreeding phase of the migratory cycle. Thus migrating birds may be geographically precise at both ends of their round trip. Such local navigational precision seems to depend on previous flight experience in the area. Certain waterfowl that pair on the wintering range are governed by the female's nesting-site fidelity. The male goes with her to breed to a site sometimes far from his own place of hatching. Pairs of some species will desert a previous habitual site and nest in a new one following an unsuccessful breeding season. Nesting-site fidelity thus ranges from close conservatism to adaptive flexibility.

World patterns

Some generalities can be drawn from the complex welter of behavior patterns shown by global bird migration. Each major part of the world has its own special features of migratory bird behavior. For example, most species of medium- to long-range travelers breed in the northern hemisphere. Many persons who live in the middle latitudes of North America, Europe, or Asia are strongly aware of this. Millions of northern birds that nest at higher latitudes stream south every fall, appearing locally for some days or weeks, and then come through again in the opposite direction each spring. As many as 95 percent of Eurasian land birds that breed between latitudes 40°N and 50°N (a band that includes Madrid, Beijing, Prague, Kiev) move south for the winter. In the northern United States and in Canada about 75 percent of the birds nesting there fly south in the fall.

One might expect an analogous northward movement in the southern hemisphere during the austral fall. But the numbers (as well as the percentages) of species involved and the distances migrated are usually minor or even trivial, compared with those of the northern hemisphere. There seems to be no instance in which *land* birds that breed in high southern latitudes in South America migrate across the

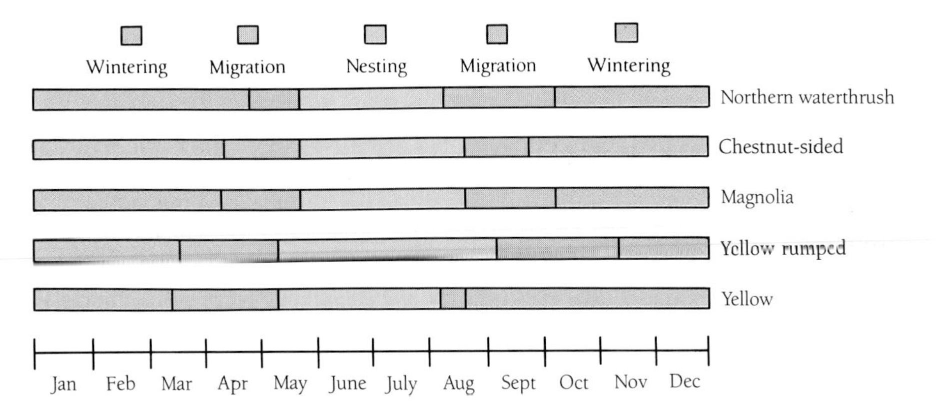

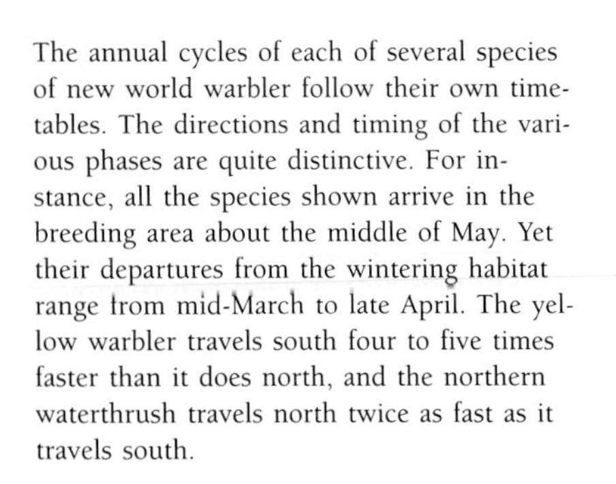
The annual cycles of each of several species of new world warbler follow their own timetables. The directions and timing of the various phases are quite distinctive. For instance, all the species shown arrive in the breeding area about the middle of May. Yet their departures from the wintering habitat range from mid-March to late April. The yellow warbler travels south four to five times faster than it does north, and the northern waterthrush travels north twice as fast as it travels south.

tropics into the northern temperate zone. Actually *transtropical* southward migration of new world land birds is not common either. Most migratory North American land birds winter in the southern United States, Mexico, Central America, and South America north of the equator. Probably the bobolink *Dolichonyx oryzivorus* is the best-known transequatorial example. It breeds widely over southern Canada and the northern United States, but winters in the pampas of southern Brazil and northern Argentina. Swallows of several species also cross the equator and range from Alaska well into temperate South America. The movements of bird migrators in the old world are quite different in frequency and extent from those in the new world. Many species of Eurasian birds on their way to wintering grounds overfly both the Sahara and the tropics into the African temperate zone even as far as South Africa. The European cuckoo *Cuculus canorus* is a well known example.

That migratory patterns of one hemisphere differ from those of the other is no doubt due to a number of factors. There is a much smaller middle-latitude landmass in the southern hemisphere. Climates in the most southern parts of Africa and Australia are quite moderate even during their austral winter. More than 13 million inhospitable square kilometers of the antarctic continent surround the south pole, whereas a somewhat greater area around the north pole is covered by the Arctic Ocean. This large mass of water, although always frozen at the pole, moderates the severity of the weather and provides a relatively rich marine habitat for invertebrates, fishes, and mammals. In contrast, Antarctica, which is perpetually covered with thick ice and snow, offers a ground habitat so hostile to life that neither land birds nor native land mammals live on the continent. All the seabirds that do live there—except the emperor penguin—migrate north to escape the harsh winter. The marine waters surrounding Antarctica are highly productive but half of the south polar zone is occupied by the nearly sterile ice cap.

In comparison to Antarctica, the arctic tundra and its adjacent forests are highly supportive of life during the brief summer. Even in winter this region harbors a surprising number of species. In northern Alaska at latitude 71°N, which is comparable to the margins of Antarctica, eleven land-bird species live in the tundra throughout the winter; twelve more are residents of nearby wooded areas. All twenty-three of these species breed in the region while at least two other species migrate only short distances within this general habitat. The summer breeding populations are swelled by thirty-five species of shore and seabirds including the arctic tern, scoters, mergansers, gulls, and four kinds of shore birds that migrate across the Pacific Ocean, among them the golden plover. In addition, more than three dozen species of North American land birds, some of them warblers and sparrows, and four species of Asian land birds fly in to breed. Included in the intercontinental visitors is the northern wheatear, a thrush whose migration is more fully described below.

Bird migration patterns in several other parts of the world are interesting to compare with those of the circumpolar regions. Australia, for example, the smallest

Enduring temperatures as low as −50°C and tempestuous winds laced with ice crystals, emperor penguins breed in midwinter on the antarctic ice shelf 100 to 200 kilometers from open water. The male incubates a single egg on top of its feet by covering it with a pouchlike fold of skin.

of the seven continents and the only temperate and tropical continent that lies completely in the southern hemisphere, has a large and diversified bird fauna (about 900 species compared with North America's 700). Yet only about six species of Australian land birds are intercontinental long-distance migrants. For some reason there are no massive migrations of small perching birds in Australia. The movements of flocks of brilliant cockatoos, parakeets, and finches tend to be nomadic or relatively short range.

Nearly 480 species (or subspecies) of European and Asian birds overwinter in Africa. About 200 of these penetrate south of the Sahara. Roughly fifty bird species fly to the dark continent from still other parts of the world: South America, Antarctica, the Caribbean, and various oceanic islands. Add to these more than 500 species and subspecies that move around within Africa itself and a pattern emerges startlingly different from that for Australia. Each migratory species seems to have its own movement pattern in which timing, directions, ranges, and locations are almost like a fingerprint.

About ninety species of birds from western and central Asia as well as around twenty-five from eastern Asia migrate to Africa, 8000 to 9000 kilometers each way even though southeast Asia, Indonesia, New Guinea, Australia, and India would seem to offer nearer potential overwintering areas. No doubt the most spectacular of these great east-west flights is carried out by the northern wheatear *Oenanthe oenanthe*. The race of this species breeding in the Alaskan tundra flies more than 10,000 kilometers *westward* across Asia to Africa. There it overwinters in nearly the same area as the Greenland race, part of which also breeds in northern Canada, but migrates *eastward* before turning south toward Africa. The combined routes of these two northern wheatear populations thus girdle most of the globe. Their travels are striking for the large range of longitude covered in addition to the more usual broad extent of latitude. It is notable that a closely related nonmigratory *Oenanthe* species breeds and spends the whole year in the African wintering area of the northern wheatear.

Effects of winds and weather

A long southward course over the western North Atlantic ocean is regularly flown by many small land birds like sparrows and warblers (passerines) on their fall migration. This was discovered by extensive tracking of birds, with mainland, island, and shipborne radar systems. Invented during World War II, radar can provide remarkably detailed data on the location, direction, altitude, and movement of flying birds or insects. In this instance the birds' nonstop flights extend from Canada and New England 2000 to 3000 kilometers to eastern Caribbean islands or South America's north coast. Piloting can be of little help to the small passerines far from land for most of this migration. How do these birds navigate?

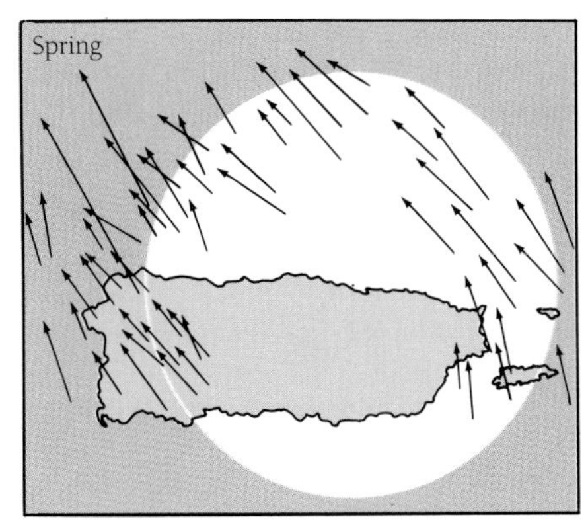

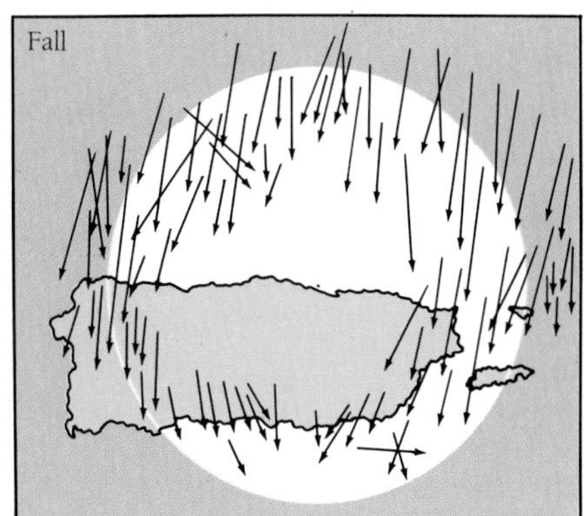

Migratory flyways like those mapped for many bird species must be pieced together from repeated careful observations made at numerous locations. Two seasonal examples of migratory flyways of small song birds and shore birds tracked overhead by radar in Puerto Rico are shown here. Each arrow shows the direction and speed of one of many flocks recorded on the radar screen. Birds observed on a spring evening in Puerto Rico were generally flying northwest toward the Bahamas and eastern continental United States. In contrast, during the fall nocturnal flights birds were mainly moving south as if they had arrived over water from maritime Canada and the northeastern United States.

Computer modeling has suggested that such long-range movements could depend on relatively simple navigation aided by strong seasonal prevailing winds. The model requires rather close timing of the birds' departure from the continent with certain repeating weather patterns common during the fall. Just after the passage of a strong cold front with its following brisk northwest winds, the birds head south-southeast out to sea in great numbers. Hence typically many of them set out together followed by periods of some days with few or no departures. If they continue to fly at their most efficient sustained speed in a direction between southeast and south they should make appropriate landfall in 3 or 4 days without further ado. Any birds that fly in too much of an easterly direction are turned toward Caribbean islands or South America by the northeast trade winds they encounter south of Bermuda. By holding a constant course out to sea and depending critically on typical tail winds to help them along, such migrators safely reach their usual destinations or way stations. Without the right weather pattern, mass disaster would ensue.

The computer model indicates that to function successfully this navigation system need not have a precise compass heading (plus or minus 15° accuracy would work), nor a specific place of leaving the continent (the birds do in fact set out anywhere between Cape Sable and Cape Henry), nor a particular flying altitude or air speed. As long as typical weather patterns prevail, a high degree of success would be expected even without fine-tuned or elaborate navigational control. The return migration of the New England and Canadian passerines in the spring follows a quite different path over the continental eastern United States and takes a month or more. Both speed and route of northward movement are quite distinct from those of the southward trip. Prevailing weather conditions and the need for repeated feeding en route during a much slower spring trip seem responsible for such path changes. Radar observations have shown fall migrations of passerines over Scandinavia and between western Europe and the United Kingdom to consist of periodic, dense waves of birds. These too are typically so timed with the weather that downwind flight carries the birds roughly in the required direction.

Wind and weather clearly affect the migration of birds and other flyers. Small weak flyers are obviously most vulnerable to these forces, but even large, strong flyers must adapt their flight behavior and navigation accordingly. The passerines just discussed take about ten times longer to fly northward in the spring as they take

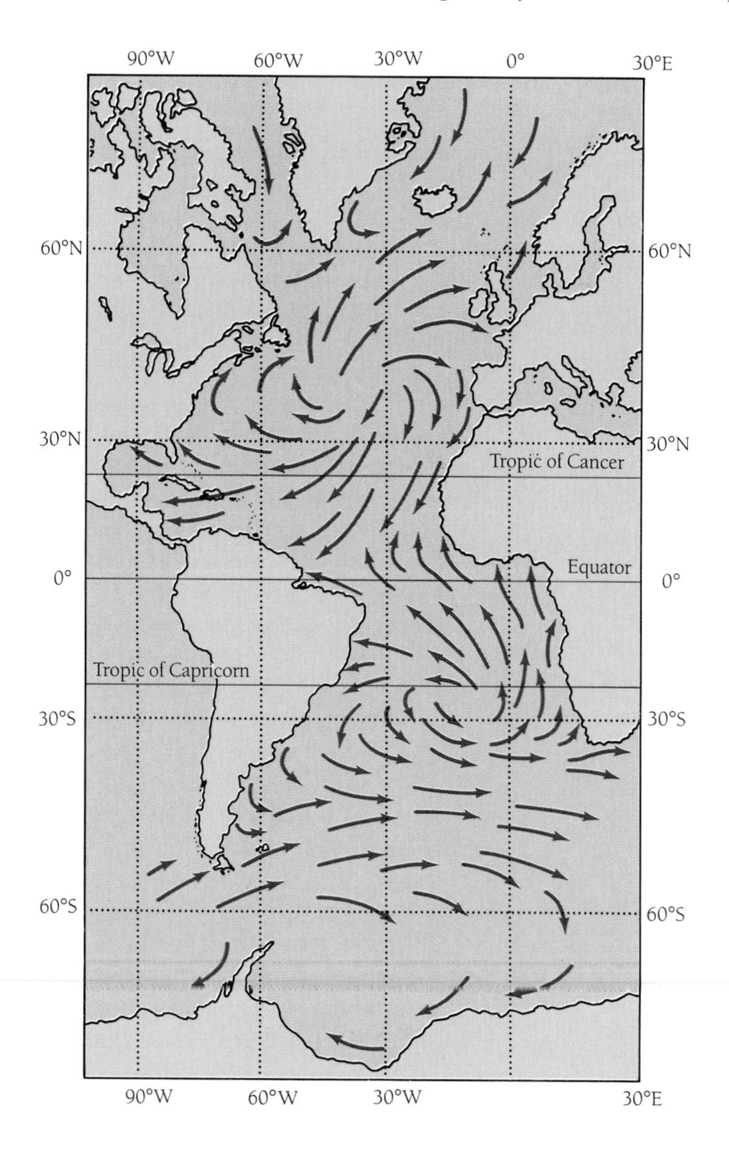

Prevailing wind patterns of the world are relatively simple over the open sea with bands of easterly or westerly winds occurring at various latitudes usually separated by intervening regions of calm or light, variable winds. However, basic wind patterns, like climate, are modified over land by mountains, deserts, and large lakes and, to a lesser extent, at sea by ocean currents. Seasons and moving weather systems also distort—sometimes intensely—the circulation of the atmosphere which produces winds. Here average wind patterns over the Atlantic Ocean during August are shown.

when wind-driven southward in the fall. Weather is, of course, notoriously unpredictable. Yet it has major reliable features that are exploited in migration. For instance, the steady westerly gales that blow at high southern latitudes propel the wandering albatross around the world in its yearly circuit. While the continents strongly modify the winds over them up to altitudes where birds normally fly, winds over the world's open oceans tend to follow a relatively simple pattern. The equatorial regions at sea, called the doldrums by sailors, typically have weak unpredictable winds. North and south of the band of the doldrums in the tropics the trade winds blow from the northeast in the northern hemisphere and southeast in the southern. At about 30° north and south of the equator are two other bands called the horse latitudes where the winds are weak and uncertain. Between 30° and 60° north southwesterly winds prevail while northwesterlies blow between 30° and 60° south. This description is quite simplified; details vary for different locations, sometimes strongly with seasons as in the tropical Indian Ocean where the monsoon winds alternate in direction semiannually.

BATS

Chiroptera (the bats) form a small compact group of specialized mammals with about 850 species, less than one-tenth those known for birds. All bats are active flyers. Chiroptera are found everywhere in the world except Antarctica. Most are tropical in distribution, but three families of insect-eating species are well represented in temperate zones. Some species hibernate, copulate, give birth, nurse their young, and roost for part of the summer all at different places in an annual migratory cycle. Roosting in sheltered sites is a characteristic feature of bat biology; usually there are different roosts for summer and winter and for day and night.

Seasonal migrations by various bats range from flights shorter than 100 kilometers to flights as long as 1500 to 2500 kilometers. Extensive banding and recovery studies in the new world and in Eurasia prove that moderate to long round trips are typical of several chiropteran families. Well documented are the massive flights of certain populations of the Mexican freetailed bat *Tadarida brasiliensis* in which hundreds of thousands of individuals fly from the southern and western United States to wintering areas in Mexico. Other populations of the same species are nonmigratory. In Europe some species have been found to migrate as far as 1600 kilometers each way. The common little brown bat *Myotis lucifugus* in New England has ranges extending to about 500 kilometers, but tree-roosting bats like those in the genus *Lasiurus* show more extensive north-south movements with the seasons. At moderate to high latitudes insect-eating bats must either seasonally hibernate or emigrate to milder climates where they can continue to feed on their prey. Many species do both, flying to lower latitudes to spend the winter resting in caves, hollow trees, or other refuges. Tropical fruit- and flower-feeding bats, like birds of the same region,

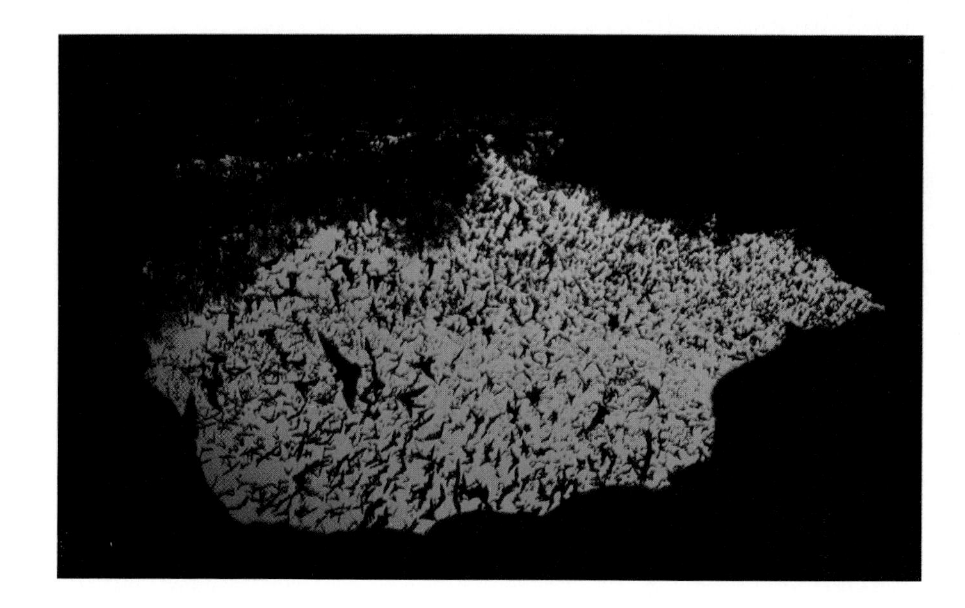

Mexican free-tailed bats emerging from Bracken Cave, Texas. This cave contains more than 20 million Mexican free-tailed bats. These bats constitute the world's largest known colony of any vertebrate animal. Tens of thousands of feet of cave walls are covered with roosting bats. Many tons of guano fall to the cave floor in summer and are mined for fertilizer during the winter, while the bats are wintering in Mexico. Each night bats from Bracken Cave cover thousands of square miles and eat a quarter of a million pounds of insects.

change their locations between the wet and dry seasons or as food availability shifts over a series of sites.

Many bats forage nightly in places 10 to 40 kilometers from the shelter used during the day. Sometimes huge flocks will stream out of the daytime roost at dusk and fly away to feed, returning again at dawn to their belfry, temple, or cave. The massive excursions of bats of the genus *Tadarida* that roost during the day in the Carlsbad Caverns are a well-known example. In addition to short-range echolocation, nighttime visual piloting is presumably used by migratory chiropterans, yet little is known about whether stars or any special geophysical techniques are involved in their navigation. As in birds, the direction of sunset may provide an important compass reference for bats. Although they are far less diverse than birds and are not known to undertake spectacular global trips, bats' migratory behavior is rather like that of a number of bird species. Differences in flight paths and timing between sexes, between individuals, between species, and between successive years are well known for chiropterans, too.

INSECTS

Although the number of species of flying insects is enormous, their life histories contain relatively few examples of long-range global movement. Among other things, this is due to their being quite small, which limits migration ranges. Even so,

Insect migration

Group	Species	Purpose	Maximum range (kilometers)
Locusts and grasshoppers (Orthoptera)	Desert locust *Locusta migratoria*	Feeding after rain	8000
Butterflies and moths (Lepidoptera)	Monarch *Danaus plexippus*	North-south, seasonal	4000
	Mourning cloak *Vanessa cardui*	North-south, seasonal	1000
Bugs (Heteroptera)	Large milkweed bug *Oncopeltus fasciatus*	North-south, seasonal	1000
Leaf hoppers (Homoptera)	Leaf hopper *Nilaparvata lugens*	East-west, seasonal	1000
Beetles (Coleoptera)	Lady bug *Coccinella sp.*	Mountain dormancy, seasonal	1000

flyers from six different orders of insects are known to migrate regularly over distances ranging from 1000 to 8000 kilometers. Two species particularly interesting for their migrations are desert locusts and monarch butterflies.

Invasions of the desert locust *Locusta migratoria* and related grasshoppers in enormous swarms have been familiar in Africa and the Near East at least since Old Testament times. Such periodic seasonal movements may range over 2000 to 3000 kilometers, and daily movements may be more than 300 kilometers. Not unexpectedly, they have been intensively studied because of their disastrous impact on agriculture during certain years. After a lull of some years, grasshoppers and several species of locusts in east Africa threatened a major outbreak in 1986. In 1988 huge populations were building up in north Africa, menacing Morocco, Algiers, and Tunisia. Like the migrations of many passerine birds, locust movements are closely linked to weather patterns. But the locusts' long-range flights are characteristically downwind, occur during the day, and follow the rains around. In central Africa, for instance, cloudlike locust swarms arrive just as the previously parched land bursts forth with nutritious shoots and leaves. To move over large distances, these insects must transform from a sedentary rather solitary phase to a migratory gregarious phase. Only individuals in the migratory phase swarm and carry out sustained flight. The mass movements are timed so that speed and direction of the wind largely determine the airborne swarm's course over the ground. Endocrine, sensory, and behavioral factors together control this remarkable behavior. The wind provides

Massive swarms of migratory locusts, like these in Ethiopia, are a recurrent plague in Africa and the Near East.

much of the transportation energy for the locusts. Even so, desert locusts frequently fly across the wind with the many individuals making up a swarm closely parallel to one another. This suggests that the animals are navigating to reach an objective.

The degree, however, to which migrating insects fly in a chosen compass direction—in contrast to choosing to fly only when the wind is favorable—is still a rather controversial point. Strong flyers such as bumblebees, wasps, and dragonflies do frequently migrate into the wind. Also nocturnal swarms of rather large insects such as moths and certain grasshoppers are often seen on radar to comprise well separated individuals not moving strictly downwind but all aimed closely in one direction. This suggests, as does the daytime behavior of the desert locust, that the animals are using some magnetic or celestial means of direction finding. On the other hand small, weak flyers like midges must depend entirely on wind to cover significant ground. But this no doubt requires them to know when the wind is right for their destination.

The North American monarch butterfly *Danaus plexippus* is one of the few insects that undertakes a really impressive long-range migration—its round trip may be as long as 8000 kilometers. As another example of behavioral diversity, its close relative the queen butterfly is nonmigratory. Even though the monarch is a common and well-studied species throughout much of North America and elsewhere in the world, entomologists have only rather recently begun to appreciate the extent and nature of its astonishing seasonal travels. It had been common knowledge for a hundred years that the population in eastern North America makes extensive treks southward near the end of summer and in the early fall. More recently these flights had been well documented by tagging butterflies' wings with extremely lightweight paper stickers and collating the recovery reports.

Many years' data of this kind clearly show that individual monarchs from southeastern Canada and the northeastern United States fly hundreds of kilometers to the south. Monarchs are day migrators resting in trees and bushes at night. Large numbers of butterflies tagged in northern areas turn up in Florida and in other Gulf States. The migration was first assumed to represent just a widespread, rather general retreat into milder areas where the frost-sensitive adult butterflies could survive the winter. Consequently, it was a stunning and beautiful surprise in 1975 when

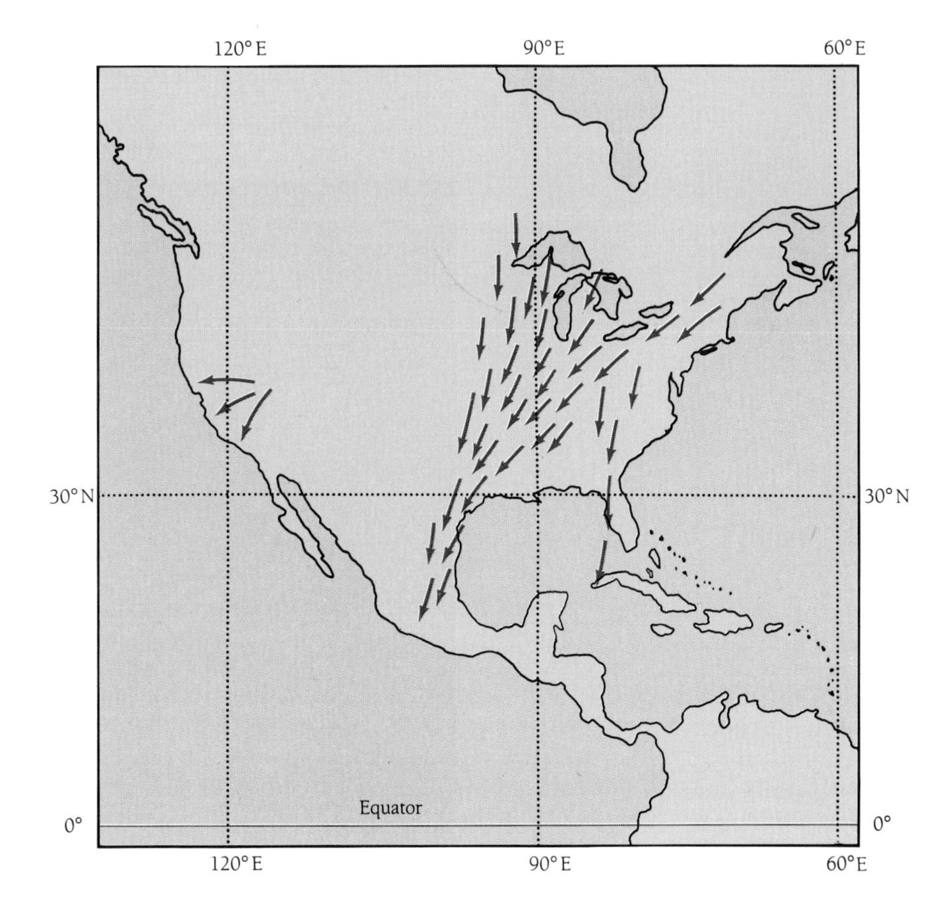

Monarch butterflies in North America have a large, broadly distributed population east of the Rocky Mountains in the northern United States and southern Canada. The frost-sensitive adults migrate south and southwest in the fall. A substantial fraction of these butterflies overwinter in sharply restricted mountain sites in southern Mexico. A smaller western population of the same species summers in the great central valleys of California and winters along the California coast from Stinson Beach, just north of San Francisco, to San Diego.

F. A. Urquhart of the University of Toronto discovered millions of overwintering monarchs in a sharply focused location at about 2800 meters elevation on the central volcanic plateau in Mexico, in northern Michoacan west of Mexico City. There are a number of particular highland areas where the butterflies overwinter. Within each of these are several specific groves or a few trees on which monarchs aggregate densely. Tagging shows that butterflies coming from summer ranges in central and north central regions of the United States and Canada overwinter in the more westerly sites of the region. Those from the east and northeast are localized more to the east. The highly favored central area where the butterfly-laden trees were first discovered holds a mix of butterflies from both summer ranges. Group site fidelity in successive years has been found to be remarkably high.

In these overwintering sites the monarchs aggregate in enormous numbers, mainly clinging to tall mountain fir trees. Functionally, they are all virgins in reproductive arrest and in a state of semitorpor. These inactive butterflies are substantially preyed upon by birds and suffer large losses if rare winter storms in the mountains are severe enough to freeze them. After 4 or 5 months of little or no activity, the still huge population of surviving butterflies once more becomes vigorous as early spring

Monarch butterflies overwintering in certain mountainous locations west of Mexico City form dense carpets on the trunks and foliage of a few particular fir trees.

warms their roosts. Their reproductive systems mature, and a vigorous mass mating ritual ensues. Thereafter both sexes begin extensive return flights to the north and northeast. Published data are scanty on how really specific the return goals are, yet suggest that, at best, they are more regional than precise. There is no evidence for highly localized site fidelity like that shown by many birds. Remarkably, the extensive, southward goal-oriented travels of the monarch butterfly are made entirely by novices! Following their return to the north the fertile long-range migrators lay eggs, thus founding the local summer population, and then die. The summer populations are then built up by two or more reproductive generations. As a result, those adults that fly south in the fall have never migrated before and have no experienced leaders. How are they able to navigate so well?

2

Animal Migration: Swimmers

Because more than 70 percent of the earth's surface is covered by oceans averaging about 4000 meters (more than 2 miles) in depth, swimming animals are able to roam the world with abandon. Also, the terrestrial 30 percent of the globe is widely accessible to such swimmers via the ponds, streams, lakes, and rivers that lace the continents even though this fresh water occupies less than 1 percent of the area covered by the seas. As an environment offering worldwide possibilities for travel, the earth's aquatic domain is second only to the air. Not surprisingly, many long-range underwater migrators are large, strong swimmers such as sharks, tunas, eels, sea turtles, and whales.

Not being aquatic, we have only limited first-hand knowledge of the underwater world. Observations from shore reveal little about marine animal behavior. Most of us never experience more than some exhilarating recreational swimming and diving. Snorkeling near the surface, relatively brief scuba-diving (perhaps down to 20 to 30 meters), as well as quite rare but more extensive excursions in bathyscaphes, submarines, and other deep-water vehicles provide humans with their only direct contact with this major part of the biosphere. Second-hand information is, of course, much more plentiful. Underwater photography and recent deep-sea television used in research have enriched our appreciation of the aquatic world. Knowledge about its animals is largely based on the practical experience of explorers, fishermen, and whalers dating back into prehistory and on the nineteenth- and twentieth-century scientific study of the seas (oceanography) and fresh waters (limnology).

Despite our handicaps for monitoring such behavior, we know that many aquatic animals migrate over distances as great as several thousand kilometers.

Sockeye salmon in the Adams River, British Columbia nearing the reproductive goal of their lifelong migratory cycle, after which they die. After several years ranging widely in the North Pacific to feed and mature these fish have returned with high specificity to the home stream where they were spawned. Chemicals in the water of their natal creek are generally believed to provide an olfactory clue for the upstream homing.

Because the world's oceans are extensive and interconnected around the poles, the longest aquatic migrations are marine or some combination of marine and freshwater. In addition, certain aquatic animals migrate within large lakes and in great river systems like the Mississippi-Missouri, the Nile, the Volga, and the Amazon.

All the animals that make long aquatic migrations are vertebrates. There is a rich and fascinating fauna of marine and freshwater invertebrates whose virtual absence from the roster of long-range migrators is explained by there being few strong sustained swimmers among them except the squids and some large crustaceans like shrimp. Squid, vigorous active predators living mainly in the open ocean, may well be found to migrate considerable distances, but our knowledge of their biology is still rather fragmentary. Many examples are known from around the world of shrimp migrating from tidal rivers and estuaries into the sea to release their eggs. After developing through early stages at sea, juvenile shrimp, along with returning adults, move into brackish water or fresh water where they mature. In temperate latitudes many completely marine crustaceans (such as lobsters and crabs) and horseshoe crabs make rather large-scale onshore-offshore seasonal migrations as do many fishes and even whales. Typically these animals live and breed during the summer in shallow inshore waters. As temperatures drop in the fall, they move out into deeper water on the continental shelf or beyond to return landward in the spring. In the Bahamas the fall treks of spiny lobsters into deeper water are sometimes spectacular for the numbers involved, but their migrations cover not more than 50 to 60 kilometers. The migratory *walk* of the wool-handed crab is probably the longest underwater journey by an invertebrate. Adult wool-handed crabs live in rivers like the Yangtze and the Rhine but migrate downstream as far as 1200 kilometers to release their ripe eggs in the sea. Although spectacular, the migrations of female landcrabs, which also must spawn in the sea, are relatively short and almost entirely over land.

Spiny lobsters *Panulirus argus* that summer in shallow water around the Bahamas move during the fall to deeper water. En route they typically queue up to "follow the leader" in columns of 20 to 60 individuals. The queueing may serve to reduce the drag of the water on each individual and thus save energy overall. Many other mobile marine animals also retreat into deeper water as the temperature of their shallow summer habitat drops in the fall.

UNDERWATER NAVIGATION

Underwater path finding is quite different from that in air largely because of constraints imposed by visibility in water and to some extent by currents. Water absorbs light several thousand times more rapidly than pure air. In addition, the scattering of light both by the water itself and by particles suspended in it degrades visibility as mist and fog do in the atmosphere. As a result even under optimal conditions, visibility underwater is extremely poor and comparable to that in a thick fog in the atmosphere. Visible sunlight penetrates only 1000 to 1200 meters into the clearest water. Hence it is completely absent—even at midday in the tropics—throughout about 70 percent of the ocean's volume. Visual piloting is thus of little use underwa-

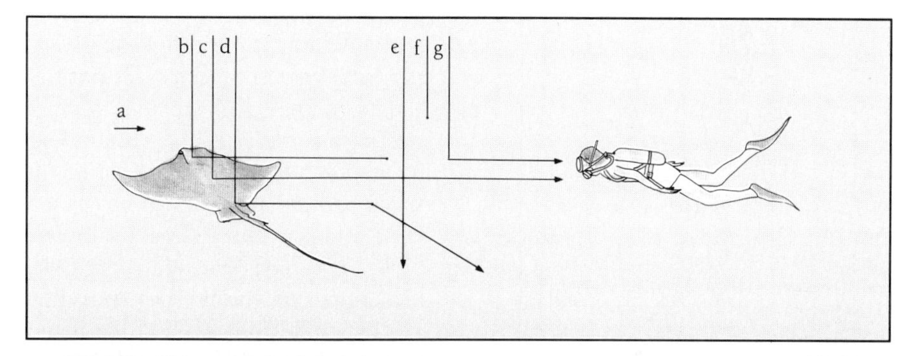

(*Top right*) To be seen in water an object must contrast perceptibly with its background. In this illustration the seven light rays all have different effects on the diver's ability to see the fish. Horizontal background light (a) determines the contrast between the object and its surroundings. Of three rays from above reflected toward the observer by the fish, only (c) contributes to visibility; (b) is absorbed by the water; and (d) is scattered out of the line of sight. Of the three rays directed downward between the fish and the diver, (g), which is scattered toward the diver, will reduce visibility. The transmitted (e) and absorbed (f) rays in this area will not affect contrast. In turbid inshore waters these factors may lower visibility ranges to less than 1 meter or even to a few centimeters. (*Bottom right*) An aquatic animal looking up near the surface of the water sees the hemisphere of the sky compressed by refraction to about half its angular extent as seen in air. The resulting bright circle called Snell's window contains the sky's zenith as its center and objects on or near the horizon around its perimeter. In this photograph a shark has been hooked by a fisherman in his boat (seen at the edge of the "window" backlighted by the sun).

ter except during the day in the upper layers and at distances of less than 100 to 200 meters at best, and often *much* less. Landmarks can be used only within this short limit and when the bottom is illuminated by light from the surface. Nevertheless, most marine animals active in the upper sunlit zone have well-developed eyes as do—surprisingly—many species such as the deep-sea rattail fishes that live near the bottom far beyond the reach of sunlight. Such highly evolved visual organs must be important in prey and predator detection and in orientation and navigation. Presumably, visual functions in the absence of daylight are possible because of bioluminescence, the production of light by organisms themselves, which is widespread in the sea.

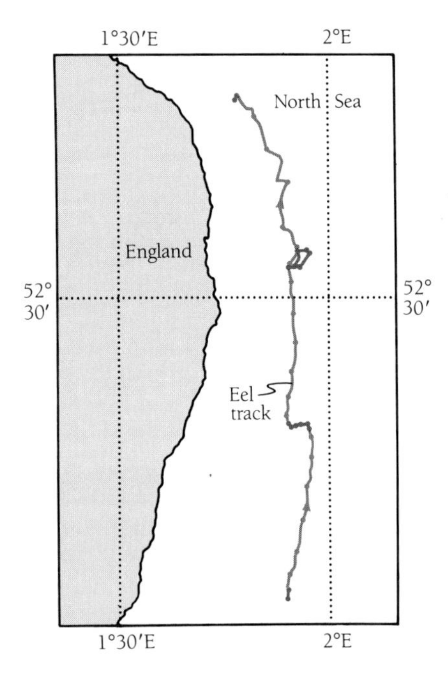

A silver eel tagged with an ultrasonic transmitter was tracked for nearly three tidal cycles (about a day and a half) in the North Sea off the English coast. During this time the fish traveled north for about 70 kilometers. It swam freely only during northgoing tides (green circles); during the period of reversed tides, the eel presumably settled near or on the bottom (blue circles). The circles record the eel's position at hourly intervals. Clearly the fish could repeatedly steer a fixed compass course over a considerable distance. It took advantage of a favorable current and avoided an unfavorable one.

Like piloting, celestial navigation is severely restricted underwater. Celestial information is only sparingly available. The sun and the moon can be perceived only within 10 to 20 meters of the surface while the stars, planets, and the sky itself can be of only minor use if any. However, the polarization pattern underwater, as in the sky, is governed by the sun's position and changes as that moves. Consequently, aquatic animals that are able to perceive it might use the polarization pattern for steering a compass course even though they are unable to see directly either the sun or the sky. Aquatic animals must therefore depend largely on means other than visual piloting and celestial navigation to control their migration. The earth's magnetic field provides a directional reference for the orientation of certain fishes. Also, underwater navigators may detect electrical, acoustic, chemical, and thermal clues for finding their way. These possibilities will be considered later.

Our terrestrialness and the severe limitations of underwater visibility hamper human efforts to understand the navigation of aquatic animals. However, much knowledge has been gained from tagging and echolocation techniques systematically applied on a large scale. Underwater, radar is virtually useless because its brief pulses of high-frequency radio waves are so strongly absorbed by the medium that they do not penetrate significantly. Sound waves, however, unlike radio waves, travel well through water. Thus acoustic systems using *sonar* do provide echolocation data on the position and movements of underwater animals. Instruments using sonar were originally invented merely as depth finders. Like radar equipment they depend on pulses of energy sent out from a ship or shore station and bounced back to a receiver by the sea bottom or by other reflecting objects. The range, bearing, depth, and movement of such sound reflections can be accurately tracked on a screen or printout. In sophisticated systems with horizontally scanning beams, such displays may closely visualize both the bottom and animals swimming at mid depths. Sonar, however, is generally unable to identify the specific kind of animal whose location in the water it readily shows. Nevertheless, these sonic systems provide an important tool for fishermen in locating fish schools and maneuvering their nets into the best position to catch them. Acoustic tracking has also proved to be a boon in studying the movements of aquatic animals. If an ultrasonic beeper is attached to a porpoise or fish, for example, the individual can then be accurately tracked for hours or perhaps days. Support for such study has been strengthened by the great commercial value of many migrating species such as sturgeon. As Harden Jones, now at the fisheries research laboratory in Tasmania, has pointed out, nine out of ten of the world's largest fisheries depend on bony fishes such as herring, anchovy, and cod that are extensive migrators.

Water currents are rather similar to winds. In general, however, they are much slower than strong winds or even breezes. But so is swimming typically much slower than flying. Even as winds help or hinder migrators on the wing, currents help and hinder the navigation and migrations of numerous kinds of aquatic animals. Winds and tides supply most of the energy for ocean currents. Winds establish the basic pattern of surface movements; tides modify the movements of inshore waters. Thus the flow of surface water in the ocean is closely coupled to the driving winds in the atmosphere. Yet typically, current velocities reach only about two percent of the driving wind's speed. The north and south equatorial currents are activated by the steady trade winds blowing from the northeast just above and from the southeast below the equator. They are separated by the equatorial countercurrent flowing from west to east in the doldrums, where the winds are generally variable and faint.

Several other factors also influence ocean currents. The distribution of continental land masses has a profound, if passive, effect on the circulation in nearby ocean areas. More subtle but highly important effects arise from variations in salinity and temperature that largely determine the density of sea water. In turn densities determine the flow pattern of deep ocean currents. The earth's rotation generates the Coriolis force that revolves water in large gyres within the major ocean basins. These current systems are primarily clockwise in the northern hemisphere where they determine, for instance, the circulation of the North Atlantic and North Pacific

The basic global pattern of ocean currents, like that of the winds, can be rather simply described at least for surface movements. As shown by the arrows in this map, the gross circulation of northern hemisphere oceans consists of huge clockwise gyres in the North Atlantic and the North Pacific. In the southern hemisphere, the South Atlantic, South Pacific, and Indian Oceans all have counterclockwise gyres.

oceans. They are counterclockwise in the southern hemisphere, setting up, for example, the basic circulation in the southern part of the Indian Ocean between Madagascar, Australia, and Antarctica as well as that in the South Atlantic and South Pacific oceans.

FISHES

Fishes are a varied and flourishing animal group that provides prime examples of long-distance movement. Ranging from jawless lampreys and hagfish, through sharks, to the many species of bony fishes or teleosts, they may be divided into strictly fresh water species and those that ordinarily live in salt water. The various habitats available to the former, ranging from tiny desert springs to brooks, rivers, ponds, and lakes, impose sharp limits on migratory routes and accessible ranges. Even so, innumerable possibilities are exploited. Among marine species there is a rather distinct break between oceanic and coastal types. The former live in or over deep water far from land. The latter live in shallow, nearshore areas, or over banks, reefs, and continental shelves. Numbered among medium to large fishes are rapid-swimming, wide-ranging tunas, swordfish, marlin, mackerel, bluefish, jacks, dolphin fish, and sharks. Some of these cover great distances in their migrations. Tunas, eels, and the salmonids will be described as models of the exceptional navigational and migratory prowess represented in this group.

Tunas

Unlike all other bony fishes, tunas are warm blooded, which affects their migratory ability. By means of a system of circulatory heat exchangers, they can raise their body temperature significantly above that of the surrounding water. This allows them to roam the high seas and into colder deep water as well as into moderately high northern and southern latitudes. Tunas swim constantly and as rapidly as ten body lengths per second. Their raised temperatures permit them to maintain the high metabolic rate necessary for such sustained, vigorous muscular activity. Their spawning, however, must be restricted to warm water because the eggs, larvae, and even juvenile fish cannot maintain the high internal temperature of fully grown adults. One of the four genera of tunas is the genus *Thunnus,* which includes three bluefin species, two in the Pacific and one in the Atlantic. All three migrate extensively; the best described of the three migrations is that of the North Atlantic bluefin *Thunnus thynnus*. This fish has a range extending across the whole North Atlantic from waters off the north coast of Brazil to those around Norway. In their annual cycle bluefins need at least three distinct subhabitats: a spawning area, a feeding area, and a wintering area. The very big individuals are the ones that undertake by far the longest migrations. Both their swimming efficiency and their ability to with-

(*Top right*) A school of rapidly swimming skipjack tuna *Katsuwonus pelamis*. Although primarily inhabiting tropical Pacific Ocean surface waters, skipjack apparently forage regularly for short times at depths of 200 to 400 meters where water temperatures are definitely subtropical. (*Bottom right*) Giant bluefin tuna (largest silhouettes) annually follow a transoceanic feeding migration that circles around the North Atlantic generally in the direction of the clockwise surface currents. The smaller and younger bluefins are found in a much smaller area.

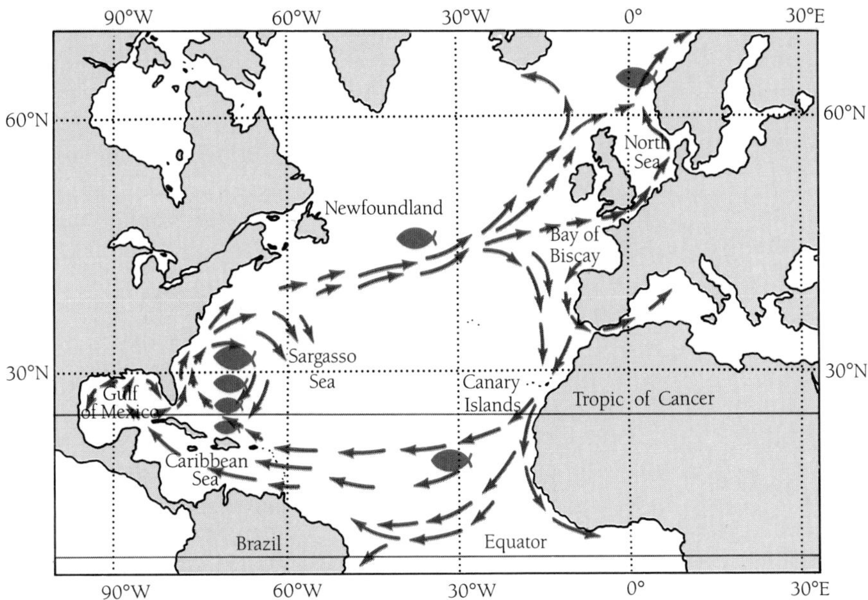

stand colder water is optimal because of their size. The smaller and younger ones, because they swim less efficiently and can tolerate less cold, are progressively more restricted by latitude and by season.

The main breeding ground for giant adults—which are more than 1.9 meters long and older than 9 years—is in the Gulf of Mexico. This tropical and subtropical area provides the warm water needed by the larvae and young but supplies little or no food for the adults. After spawning during May and June, the sexually spent, lean, and hungry fish, following the Gulf Stream, migrate rapidly north to the coast of New England and Canada's Maritime Provinces where they feed voraciously. In July and August part of this big-adult population, still tracking the clockwise circulation of the North Atlantic, swims east-northeast across the Atlantic Ocean. Some go as far north in search of food as the coast of Scandinavia; others enter the North Sea through the English Channel. The one-way swim from the Gulf of Mexico through the Straits of Florida to Cape Cod and on to North Cape covers 7500 kilometers or more!

By September and October these large tuna, now well fed and fat, begin to migrate south from the North Sea and Norwegian coast to wintering grounds near the Canary Islands where they are found in November and December. At this time the smaller, less venturesome individuals that remained behind in the western North Atlantic, are on their way south to Caribbean wintering grounds. The larger eastern Atlantic stock move next from the Canaries southwest and west across the Atlantic towards the Brazilian coast. Hence the overall annual migration pattern is a downstream circuit of the oceanic current system. The fish continue to follow this circuit toward the Caribbean during March and April, trailing the western group of giants, which did not go on to Norway and the Canaries, into the Gulf of Mexico. Both groups breed there in the next two months, starting the cycle again. When younger fish in the first and succeeding years of development are included, the migratory pattern is even more complex. Altogether then, this species vigorously exploits much of the North Atlantic as its domain.

Diadromous migrators

Most fish, like the tunas, are strictly limited in the choice of their migratory routes by the saltwater-freshwater boundary between seas and rivers. Some marine species, however, are more versatile and can enter estuaries or other areas where salt water is diluted by fresh water. Some are even able to ascend rivers into fully fresh water. Conversely, there are freshwater fish that can also flourish in estuaries or even fully marine habitats. Some species spend extended periods functioning well successively as freshwater and as marine animals. Fish that can migrate back and forth between the two physiologically quite distinctive environments are called *diadromous*. In New Zealand, for instance, seventeen of the twenty-seven freshwater species are in this category. The best known of diadromous fish are the eel and the salmon. They both spawn and spend part of their juvenile life in one medium but migrate into the other to feed and grow. Yet relative to each other their life cycles are inverted. The eels breed in midocean, the salmon in freshwater streams. These particular fish movements differ sharply in one respect from those of most small birds, which reproduce and migrate annually: Growth of Pacific salmon and anguillid eels to maturity requires several years and only one migratory cycle is completed in the fish's lifetime. Little scope seems available here for learning and experience. Still more diversity of this sort is found in various sea birds and whales that may require as long as 5 to 25 years to reach sexual maturity during which they may in fact move about extensively. Thereafter they migrate and reproduce regularly at 1- to 3-year intervals usually characteristic of the species.

Eels

Fish in the worldwide genus *Anguilla* spawn far from land quite deep in tropical or subtropical ocean basins. Rather high water pressures like those in deep water seem to be needed among other things, to bring about eel reproduction. The weakly swimming, ribbonlike larvae travel thousands of kilometers from pelagic breeding zones in the Pacific, Indian, and Atlantic oceans to enter freshwater streams as small glass eels. Some years later these same fish at maturity descend to the river mouths whence they swim back to their remote breeding grounds. This moving downstream to spawn in the sea, called *catadromous* migration, is notably less widespread among fishes than is moving from the ocean to spawn in fresh water. However, as already mentioned, shrimp in many parts of the world as well as the wool-handed crabs of east Asia and western Europe move from streams to seas to spawn. To look at eel life cycles in more detail, consider the two closely similar North Atlantic eels: *Anguilla rostrata* (the American eel) and *Anguilla anguilla* (the European eel). Both breed in the Sargasso Sea south of Bermuda, with some overlap in spawning areas. American eel juveniles and adults live in fresh waters ranging from Greenland and Labrador, south to Mexico, and into some regions of northern South America; juvenile and immature adult European eels range in fresh water from Iceland to Scandinavia, southward in western Europe to rivers flowing into the Mediterranean and the Black seas, and to the Canary Islands and Morocco.

Eel migratory routes in the Atlantic Ocean differ not only between the species but also between the willow-leaf-shaped *leptocephalus* larvae, which travel to fresh water, and the large mature adults, which return to the Sargasso Sea. Most of the larval movement is thought to be passive and determined by prevailing currents. Drifting together to begin with in the Florida Current and the Gulf Stream, larvae of both species travel at rates roughly predicted by current velocities. American eel leptocephali leave their oceanic "escalator" and arrive at their eastern North American freshwater destinations after 1 or 2 years in transit. Their European cousins stay with the currents for 2 or 3 years, crossing the Atlantic from west to east in a journey that may cover 6000 kilometers from spawning area to feeding ground. As they arrive at continental river mouths, eel larvae metamorphose into juveniles, called glass eels, or elvers, which actively migrate upstream for long periods of feeding and growing.

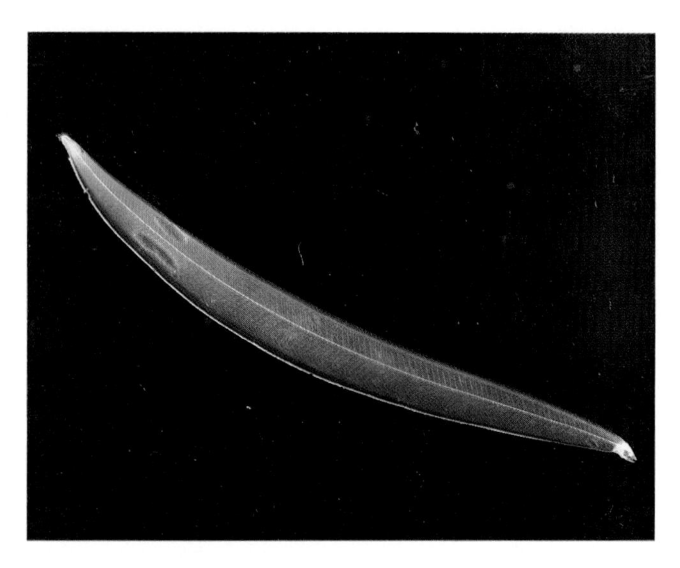

(*Top left*) Larval eels differ so strongly from their postlarval stages that they were for some time incorrectly classified. Usually thought to be feeble swimmers, their distribution to North America and Europe from their hatching place in the Sargasso Sea is mainly attributed to passive transport by the Gulf Stream. This preserved specimen shows the flattened willow-leaf-shaped form of the larva, which is remarkable for lacking hemoglobin in its blood. (*Top right*) After drifting with the clockwise currents, the leptocephalus larva on reaching the brackish water of a river mouth metamorphoses into a transparent eel-shaped fish—the glass-eel, or elver. These migrate upstream to feed and grow for many years in fresh water.

After as many as 15 to 20 rather sedentary years in fresh water, the large, yellow eels move downstream to the river mouth. From there, transformed into seagoing silver eels, the sexually ripening adults swim to the breeding areas in the Sargasso Sea by a more direct route than they took earlier as drifting larvae. The convergence of these two *Anguilla* species on a single remote spot to breed poses a mystery. Because the freshwater habitats of the eels are spread over an enormous geographical area, many distinct return courses are traveled; in human terms each would require a different set of compass headings. Obviously, even an individual eel usually follows quite different routes and has different navigational needs on its eastward and westward crossings of the North Atlantic.

The extraordinary pelagic phases of this circuit were first proposed by the Danish fisheries biologist Johannes Schmidt in the early 1920s. By studying the distribution of larval eels in detail, he found that the smallest and youngest stages were found only in quite restricted areas of the ocean south of Bermuda. The spawning sites, he reasoned, must be there or close by. Such a breeding place seems particularly strange because of the far-flung distribution of the freshwater eels. Indeed doubt has been cast more than once on Schmidt's hypothesis. Particularly hard to believe for some is the notion that larvae spawned in the south central North Atlantic could find and enter the Mediterranean on their way, for instance, to the Nile. Equally astonishing to the doubters would be the return trip by the same but now mature fish from such a distant river back to the Sargasso Sea. One alternative is that eels from Mediterranean rivers and tributaries breed within that inland sea itself. D.W. Tucker rejected this possibility but also rejected Schmidt's notion that Euro-

Large yellowish-brown eels live in fresh water for up to two decades before migrating downstream to the river mouth. There they are transformed into an adult seagoing form that develops a resplendent silver color and ripening gonads. Their eyes and olfactory organs grow much larger to cope with the problems of long-range underwater navigation, and at the same time their digestive tracts degenerate. The silver eels do not feed on the long trip to the Sargasso Sea, where they die after spawning.

pean adult eels undertake a return breeding migration to the Sargasso Sea. Instead he argued in 1959 that this eel is of the same genetic stock as the American one. In this view European eels, including those in the Mediterranean, are individuals that will never breed because they drifted too far from the breeding grounds. The very existence of European eels would thus have to depend on a continuing supply of drifting larvae of North American parentage. Whatever its other merits may be, Tucker's iconoclastic view evoked a wave of fresh interest in the "eel problem." So far the accumulating evidence appears to be coming out against Tucker's interpretation.

Sparked by F.W. Tesch of Hamburg renewed field research to resolve these conflicting issues has largely confirmed Schmidt's main postulates. Sixty years later the youngest leptocephali of both species are still found in subtropical oceanic locations, somewhat larger in area than the Danish pioneer had originally claimed, but otherwise similar. Mediterranean spawning of the European eel has never been confirmed. Furthermore, extensive biochemical "fingerprinting" has shown that the American and European eels are of different genetic stocks. Despite such interesting new work, some puzzling questions remain unanswered. For instance, are the eastern and western North Atlantic eels sufficiently different to be two species, as they are often named, or are they merely distinct races of one? The critical evidence on whether they normally can interbreed is not available. Considerable fishing effort in the Sargasso Sea has failed to capture any intact pre- *or* post-spawning adult *Anguilla* or their shed eggs. In addition almost no silver eels have ever been caught anywhere in the ocean after they have left shallow continental waters such as the Baltic! Some progress has been made, again led by Tesch, through tracking sonic-labeled silver eels as they move from river mouths and estuaries over the continental shelf into the open Atlantic.

Salmon

The salmonids are a family of teleost fishes that includes salmon, trout, and char. Many of them are long-range migrators and expert navigators. These are all *anadromous* fish that ascend rivers to breed, as do river lamprey, sturgeon, alewife, shad, smelt, and striped bass. All salmonid species breed in the fresh waters of clear rivers, lakes, or their rapidly flowing tributaries. Typically, immature fish descend their natal streams to spend part of their life in the sea. This migratory urge in various populations ranges from "sometimes" in rainbow trout to "usually" in coho salmon to "always" in chinook salmon. The salmonid marine phase lasts 1 to 5 years and ends with the precise return of maturing fish to their home streams where they spawn. The full life cycle is made up of four phases, each of which may require different behaviors and navigational skills.

During the *first phase* juvenile salmon live in fresh water. Usually the eggs from which they hatch are laid and fertilized in shallow, fast-flowing water. The eggs are sheltered from the current between the stones of gravel nests excavated by the

Salmon farmers in North America, following successful developments along Norway's arctic coast, have recently learned to cultivate both Atlantic and Pacific species with commercial success. Here market-sized salmon are harvested on a farm on Grand Manan Island, New Brunswick, Canada.

female. Rapid currents pose a threat also to newly hatched fry and juveniles. They must orient themselves against the current and swim at least as fast as the water is flowing or they will be carried away. At the right time the young fish do move downstream toward the mouth of their river. Before migrating from fresh water to the sea the juvenile salmon of most species transform into a *smolt*. Its body turns silver, as do those of seagoing eels, and changes structurally and functionally as it becomes adapted to life in seawater. Pink and chum salmon may take only hours, or at most days, to move downstream to the sea. Other species of Pacific salmon take much longer (often 1 or 2 years). Sea trout, especially in high-latitude northern rivers, may not go to sea until they are 5 years old. The Atlantic salmon sometimes spends as long as 7 years in fresh water before venturing into the ocean. Once in the sea cutthroat and sea trout inhabit inshore waters close to their river mouth. In contrast, the Atlantic salmon and the Pacific sockeye journey hundreds or thousands of kilometers from their rivers. During the 1 to 5 years salmon spend in the ocean, they feed extensively and achieve most of their growth. The navigational objectives of this *second phase* must include finding suitable food as well as avoiding predators, both particularly critical for small juveniles.

Accordingly, the outward marine migration of salmon would appear to require well-regulated, nonrandom behavior. This activity is in fact much better documented than the teasing case of the silver eel. Whereas almost no eels have ever been fished in the ocean, salmon are caught commercially by the thousands in both the North Atlantic and the North Pacific. Data gathered by extensive tagging projects over many years tell us quite a lot about these fishes' marine phase. In leaving rivers for their estuaries, smolts have been observed to go out with the tides. Ocean currents, too, may play a role in reducing the cost of locomotion as do tail winds for

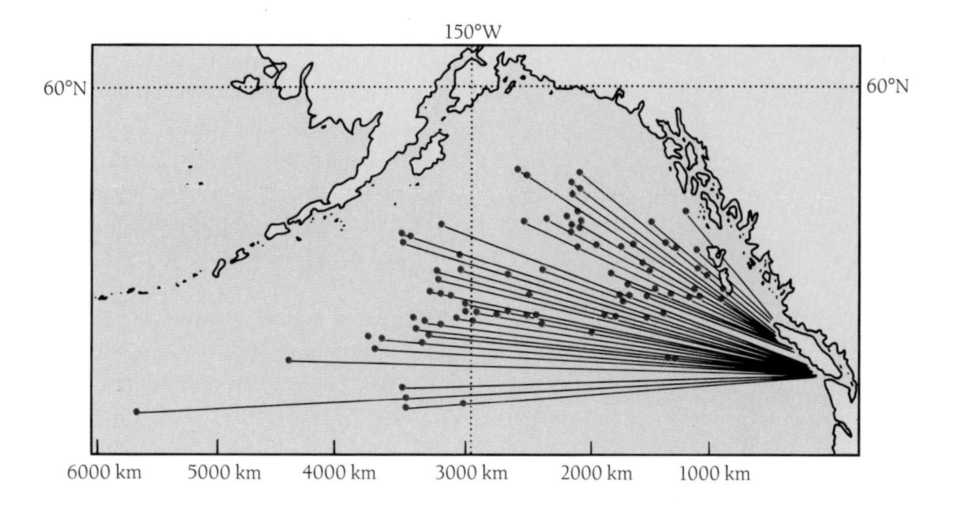

Oceanic homing behavior of maturing sockeye salmon caught and tagged at sea (red dots) and recaptured in the Vancouver Island area en route to the Fraser River. Only 373 fish first caught at least 1000 kilometers from the river mouth and traveling faster than 45 kilometers per day are shown. Calculations indicate that these fish were probably swimming in relatively straight lines day and night to account for the distances and time recorded.

some birds and insects. However, such water flow is much slower than air currents and even far slower than currents in most rivers. Consequently crosscurrent swimming and transfer from one type of water to another are general components of these open-sea movements. Populations of the various Pacific salmon species mingle in this phase of their life cycles. Stocks from the west (e.g., Japan and the Soviet Union), from the north (e.g., Bristol Bay and Kodiak, Alaska) as well as from the east (e.g., the Fraser River of British Columbia and the Columbia River) mix together in the North Pacific south of Alaska and the Aleutians.

In their *third phase* salmon return as maturing adults to the mouths of their native rivers. The intermixed species and populations sort themselves out and home successfully, often over thousands of kilometers. We know that the navigational behavior of these fish *is* tied to particular geographic objectives because individual fish return exactly to their parents' spawning site. Home-stream fidelity between 84 and 98 percent has been reported for recaptured wild populations of various species and locations. Extensive breeding experiments clearly show that precise homing depends on genetic factors as well as juvenile learning.

Nevertheless, as Harden Jones has pointed out, when individuals of three species of Pacific salmon venture into a certain south-central part of their normal oceanic ranges, they seem never to get home! This conclusion was drawn from a detailed analysis of thousands of tag returns of fish marked on the high seas and recaptured in coastal waters. Salmon tagged in an area centered on the international date line, and south of 48°N latitude, were not recovered within a year in coastal waters anywhere! The normal navigation techniques seem to fail in a particular part of the ocean. If this is true, future study of this no-return phenomenon may well help us understand how successful migrators do get home.

Sperm whales, although they are air breathers, dive deeply and long in foraging for squid, a major item in their diet. Adult males migrate annually from tropical and temperate breeding areas to circumpolar fishing places. Like porpoises and killer whales, sperm whales are toothed whales but have their large conical teeth only in the narrow lower jaw.

Extensive introductions of salmon into the southern hemisphere mostly failed to establish self-reproducing, seagoing stocks. One exception is found in a central region of the east coast of South Island, New Zealand. Certain rivers there do now have substantial breeding stocks of quinnat salmon, which is the same species as the North Pacific chinook salmon. Attempts to explain repeated transplantation failures of both Atlantic and Pacific species to other parts of New Zealand, as well as to Tasmania, South Australia, and the Falkland Islands, first proposed that to navigate effectively at sea the fish need suitable ocean currents and gyres. More recent work in the Pacific Northwest, however, suggests that the explanation may be more general and biological. Natural selection has apparently molded the genetics of individual salmon populations so distinctly that transfer even to nearby rivers usually decreases reproductive and migrational success even though returns from the sea to the home estuary may seem normal.

For those salmon that do not get lost but rather reach their home estuary, the *fourth phase,* the upstream migration to the spawning area begins. This is the second period of the life cycle spent in fresh water. The would-be breeders must swim successfully against the current for various distances—rather short for the pink salmon but, as great as 3000 to 4000 kilometers for others ascending a long river like the Yukon. At every fork in the river the correct choice must be made. While some details are still quite controversial, it seems certain that *chemical clues* in the water permit a salmon not only to recognize its home estuary but also to make all the correct choices to reach the very spawning area where it hatched. Pacific salmon die after spawning. Thus there are no teachers to show fry and juveniles how and when to migrate. Trout and char do breed repeatedly, as do Atlantic salmon to some extent.

WHALES

Another remarkable group of aquatic migrators are the whales. Cetaceans, which include porpoises and whales, have aroused extraordinary public interest in their size, songs, and migrations. As descendants of four-footed, air-breathing, terrestrial mammals, they represent a notable reversion to the fully aquatic existence of early vertebrates. In making this reversion, whales turned around the evolutionary invasion of land pioneered by amphibians perhaps 200 million years before! Even so, cetaceans remain strictly dependent on frequent returns to the surface to breathe, no matter how far or deep they swim and dive. Because they are highly proficient swimmers whose size ranges from rather large to gigantic, we would expect these marine mammals to be capable long-range migrators. Unfortunately, most of our knowledge of cetacean migrations depends either on data for species like the gray whale that are easily observed from well-populated shores or on information from the whaling industry. The latter collects data only on commercially valuable species, usually reporting only when and where individual animals were killed. Consequently, scientists know very little about the migrations of the great majority of the seventy-five or more cetacean species, including the numerous kinds of porpoise.

The sperm whale *Physeter catodon*—a partial migrator—has a complex social structure based on age, sex, and season. Apparently females and immature individuals remain in temperate or tropical latitudes all year. Organized in mixed schools, they are joined by large bulls in the mating season. Young maturing males later form independent bachelor groups that persist for some time. Whether these male schools or the mixed ones have any fixed pattern of oceanic movements is not certain. However, outside the breeding season full-grown males do undertake large-scale migrations north or south to circumpolar seas. In contrast to many migratory birds, the male sperm whale breeds in the tropics or temperate low latitudes and uses polar regions as feeding grounds. A number of other whale species follow this geographic pattern.

Both sexes of baleen whales migrate. Lacking teeth in either jaw, whales of this group have a curtain of finely divided "whalebone," or baleen, hanging down from the upper jaw. Acting as a sieve, the baleen strains out small food organisms, such as the Antarctic krill. To feed successfully these whales must swim through highly productive, plankton-rich waters with their mouths wide open. The baleen whales include the humpback, right whale, fin whale, sei, minke, and blue whale. The last of these (*Balaenoptera musculus*), a leviathan that may be 30 meters long, is often described as the largest animal that has ever lived on earth, including the dinosaurs.

(*Below*) Propelled by its large tongue, the baleen whale's food is strained out of mouthfuls of water by a curtain of fine strips of whalebone (baleen) hanging down from the upper jaw. Close up this filter in the gray whale looks rather like a giant toothbrush. (*Below right*) The relation of the gray whale's baleen to its mouth and head is seen here in an animal poking its snout up out of the water. The upper jaw with its curtain of ivory-colored baleen is uppermost to the right. Both jaws and the top of the whale's head are heavily encrusted with barnacles and whale lice.

A mother humpback whale near the sea surface with her calf, which will remain dependent on her for about a year.

All baleen whales forage at high latitudes for about 4 months and then feed little, if at all, for the rest of the year, when they are away from the Arctic or Antarctic. This remarkable behavior pattern may be dictated by the need to calve in subtropical or tropical waters where the cold-sensitive newborn can survive. But tropical seas generally contain only about one-tenth as much food as is available to these filter-feeding giants in their richly productive high-latitude subhabitat. Clearly the benefits of warm-water calving must balance the costs of 8 months of near fasting.

The humpback whale *Megalopterus novaeangliae,* although cosmopolitan, has distinct North Atlantic, North Pacific, and Antarctic populations. Recent photographic identification of individuals by fluke and body color patterns has yielded striking information on their large-scale movements. For instance, particular individuals that feed together in the Gulf of Alaska apparently migrate to different winter breeding areas. Some go to Hawaii, others to tropical islands off the coast of Mexico. The western North Atlantic population feeds as far north as the coast of Greenland in the summer and migrates to specific winter breeding areas in the West Indies between the Virgin Islands and South America. Fidelity to particular high-latitude feeding sites has also been documented.

The gray whale's long migratory journey extends annually from arctic seas through the Bering Sea, Bristol Bay, across the Aleutian chain, and close along the continental shore as far as subtropical lagoons on the ocean side of Baja California, where mating and calving take place in successive years.

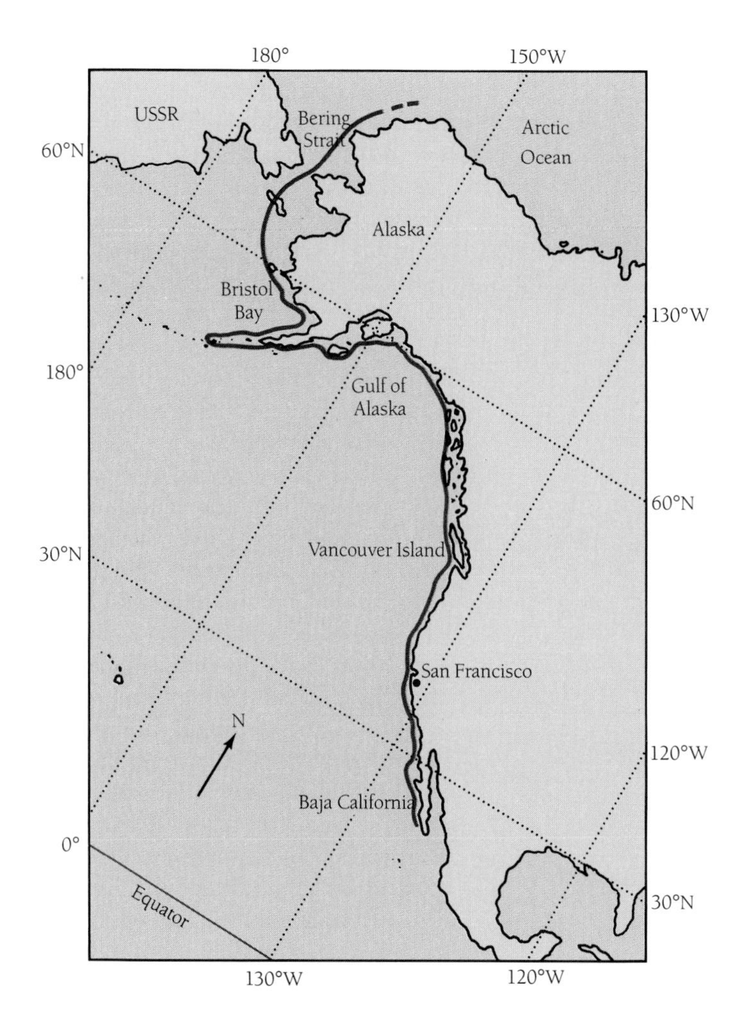

Unlike all other cetacean filter feeders, the gray whale *Eschrichtius robustus* feeds primarily on the bottom, sucking up great patches of algae and soft mud to strain out small crustaceans. This baleen whale currently lives only in the northeast Pacific but until the end of the nineteenth century it was also common near Korea and Japan. It apparently lived in the western North Atlantic until the seventeenth century and near western Europe before that. Its major foraging areas today extend from well above the arctic circle near Point Barrow on the Beaufort Sea coast of Alaska to the Chuckchi and Bering seas. From there the gray whale travels as far as 10,000 kilometers to breed in subtropical Baja California in Mexico. This is the record for such a journey by a mammal.

The migrators from the northern part of the range move in early winter through the Aleutian Islands to follow the southern Alaska coast closely. They continue to swim close to shore past British Columbia, Vancouver Island, Washington, Oregon, and the whole length of California. A few shallow lagoons on the Pacific side of Baja California are the objective of this long trip. Here is another case for which we know for sure that certain individuals return to precise locations year after year. Their navigation is sharply and effectively goal oriented. First arrivals appear in the lagoon in January but peak numbers arrive in the first half of February. There in the calm, warm, protected areas, the gray whales calve and nurse their young. The cool waters of the summer feeding areas and even the warmer waters of the California current are too cold for baby whales until their body temperature regulation develops. Gray whales also court and mate in Baja; each year some females that were not gestating as they swam south become pregnant while there. During her life-span of perhaps 20 to 30 years, a female gray whale may produce a dozen offspring, one at a time. These and other long-lived animal migrators may repeat the same trip many times; repetition undoubtedly sharpens their navigational skills. Even a simple social structure can help regulate migratory behavior in schooling fish and flocking birds. Whales are significantly more organized than that; they care for their young, communicate with each other, and form pods by sex and size—all of which may help individuals learn and execute their navigational tasks. Such interactions transfer information by nongenetic means, an important feature of social evolution.

Gray whales migrate northward in the spring. First, in February and March, all whales except mothers with calves follow a broad offshore course. They generally travel in straight lines from one promontory or cape to the next. Juveniles swim closer inshore than mature whales. The cows and their calves migrate 2 months later. In this case the inexperienced young can learn to navigate from their veteran mothers. At least in fair weather and during the day, gray whales travel mainly within sight of the mainland or islands, sometimes even in the surf. Consequently, landmarks observable above the water could provide guidance for piloting. Like several other species of whales, grays are known to stick their heads out of water as if reconnoitering. Could such "spy-hopping" be evidence for the use of visual landmarks?

MIGRATION AND NAVIGATION

What have we learned about animal navigation from our review of long-range migrations in Chapters 1 and 2? For many species migratory skill is often inherent. Insect, fish, and bird migrators carry out global goal-achieving movements without previous experience or the help of experienced individuals. This proves that at least several elements of navigation, such as when to start and which compass heading to follow, must be genetically programmed. However, repetition increases the speed

Gray whales follow the coastlines closely on their long journeys between the Arctic and the subtropics of Mexico. Consequently "spy hopping" such as that seen here could be part of their navigation.

and accuracy of the trip in some cases. It also can significantly improve an animal's ability to cope with unpredictable disturbances like bad weather or experimental displacement.

Exploration, repetition, and memory are clearly important for migratory behavior. Just as clearly, they are not its only elements. For instance, migration in many instances is a social phenomenon in which enormous schools or flocks participate. This may permit the most experienced or most perceptive individuals in the group to lead or modify its collective activity. In various examples navigation is clearly possible day or night, over land or sea, including trackless deserts and great expanses of open ocean. One of the most impressive features of animal migrations is their fantastic variety. Every species and population has its own geographic track and detailed timing. This surely means that whatever navigational techniques are used, they must be quite flexible as well as versatile. Before analyzing these mechanisms further in animals, let us briefly consider how humans cope with comparable problems of finding their way.

3

Human Navigation

Humans have been ranging widely over the face of the earth since modern *Homo sapiens* evolved. For example, successive waves of human immigrants have crossed the vast Pacific Ocean over the last several millenia. Starting from southeast Asia and Australia, the ancestors of several present-day human groups, the Austronesians, set off eastward, exploring and settling the many thousands of islands dotting the world's largest ocean. Finding new places to live, fishing, trading, and waging war gave impetus to bold seafaring. As time passed and the ethnic groups we recognize today evolved, the Polynesians proved to be the most adventurous, ultimately reaching their limits in Hawaii, Easter Island, and New Zealand. They also sailed eastward on the Indian Ocean as far as Madagascar. They apparently visited even the Galápagos and South America but did not settle. These long voyages were carried out completely without magnetic compasses, maps, and clocks. Accident and good luck may serve to discover a previously unknown destination but deliberate returns home or repeated round trips require effective navigation. As the Polynesians did return home and did make round trips, their travels, like the migrations of many fish, birds, whales, and other animals, surely depended on navigation. If we could find out how they navigated, our task of understanding the way animals go from one place to another might be considerably advanced.

INTERISLAND TRAVEL

A traditional Caroline Islands navigator plying his craft.

In fact we can study the traditional methods of human pilots in a surviving ancient society. Fortunately, and rather amazingly, there still remain in certain small remote Pacific islands, vestiges of essentially stone-age navigational methods. Like so-called living fossils, such as the horseshoe crab, these persistent traces of early human skill

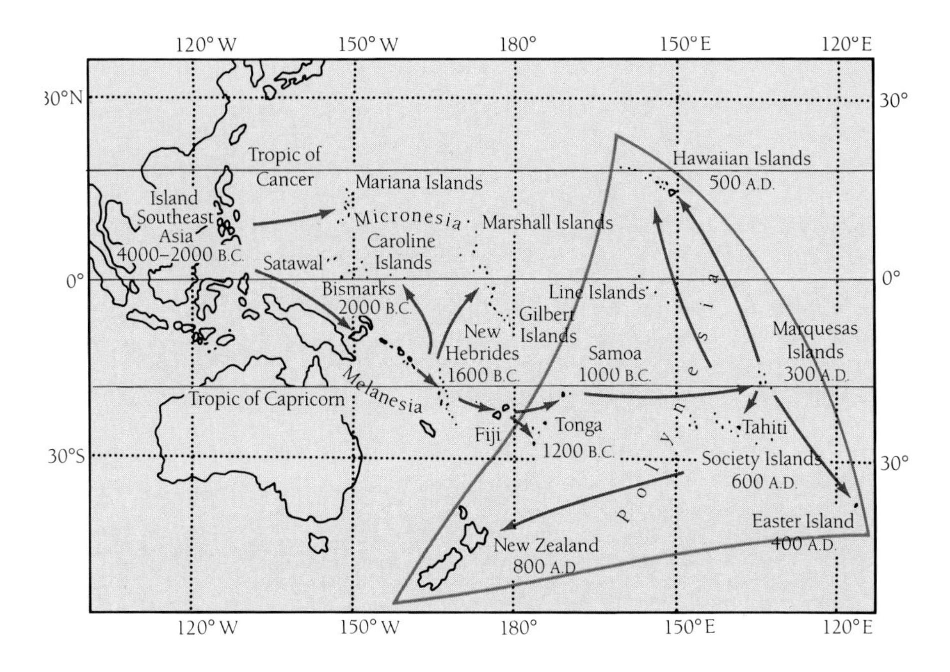

Starting from Southeast Asia between 4000 and 2000 B.C., the settlement of the Pacific islands from west to east continued for several thousand years. The map includes dates when human settlers probably arrived on various islands. Easter Island, the Hawaiian Islands, and New Zealand, which define the Polynesian triangle, can be seen as remote, isolated, "end-of-the-road" places.

are there for our instruction and understanding. To be able to observe and be taught by living practitioners of a prehistoric mariner's art is a heady experience to savor in the late twentieth century. From them we can learn that chance and luck are not the only elements in their sailing exploits.

In the four centuries since Europeans first heard of them, canoe voyages by Polynesians and Micronesians have become far more restricted. Present-day routine journeys by these Pacific people are not longer than 150 to 300 kilometers. Yet in relatively recent times regular round-trip voyages over distances of 650 to 850 kilometers, for instance between Caroline atolls and Saipan, have been repeatedly recounted with enough details to make them convincing. Long ago, however, a number of 6600-kilometer round trips between Tahiti and Hawaii were made, apparently in the twelfth to fourteenth centuries. Recently a Micronesian navigator using only his traditional skills sailed successfully from Hawaii to Tahiti. His vessel, the Hokule'a, is a modern replica of a 20 meter, double-hulled canoe built from plans recorded by Captain Cook for an actual eighteenth-century Polynesian ship. This voyage shows that the effectiveness of traditional navigation is not limited to local well-known trips. In addition a pilot adept at this technique can reach a specified distant location to which he has never traveled before.

A few recent reconstructions like this of great sailing canoes (15 to 30 meters long overall) have served to recall a once great maritime tradition. However, the best information on the old methods of navigation comes from a few enthusiastic researchers who, following earlier explorers, have combed the South Seas for still practicing native navigators and have recounted for us their dedication and skill at interisland sailing. For instance, both Thomas Gladwin and David Lewis have independently described how they learned Micronesian and Polynesian ways of navigating from practicing experts. Tutoring, along with extensive hands-on demonstration, provided their instruction in the Central Carolines, the Santa Cruz Islands northwest of Fiji, Tonga, the Gilberts, and the Marshalls. Differing in certain details, the sailing methods in the various island groups nevertheless show enough similarity to suggest an ancient shared origin. Indeed Lewis concludes that the Pacific island navigation system he has studied does not appear to have changed much from those of the Indians and Persians, the Chinese, and probably even the Phoenecians, splendid navigators three millennia ago. The broadening recent interest in these traditional sailing techniques apparently has stimulated something of a local renaissance in native canoe travel and navigation in the central Caroline Islands.

Traditional navigation is an art for which aptitude and long training, including painstaking practice at sea, are absolutely essential. The facts that must be learned, such as the locations relative to home of perhaps 50 to 100 islands, the times and directions of the rising and setting of the sun plus those of many different stars, imposes a heavy load on the would-be pilot's memory. The noninstrumental navigator must be incredibly aware and keen to con every possible clue in nature around him (the waves, the clouds, the sky, the winds, the fish, the birds, and so on) and to combine them with his vast store of remembered lore. Much of the essential detail is scarcely detectable at first to western observers even when they themselves are experienced and skillful instrumental navigators. As might be expected, the teaching of all this, both by rote, myth, and chant, and by extensive practical experience, is in some places from father to son or nephew or from grandfather to grandson. In the old days on some atolls group teaching was done in a canoe house by village elders. The acknowledged great pilots (and teachers) on an atoll or in an island group even today are highly regarded VIPs ranking with, or outranking, the chiefs.

Marshall Islanders use stick charts like this one as teaching aids to train navigators. Small cowrie shells represent islands, while coconut palm ribs tied together with plant fibers represent major wave patterns.

Although these pilots use no instruments in the navigation itself, they do use teaching aids in training for the job. They make stick charts visualizing wave patterns around particular islands but do not employ them at sea as we do maps. Instead these charts serve the Gilbert and Marshall island navigator-teachers for instructing their pupils on land and also, perhaps, for storing traditional information. Pebbles and sticks, or patterns of stones on the ground, are sometimes used to represent islands or stars. In the Gilbert Islands ceiling beams and roof-thatch areas seen from a given location in the canoe house provide a grid for specifying star positions and sky quarters.

Experienced Caroline Islands navigators provide training on land with the aid of stone "maps."

Navigational methods

Contemporary Micronesian and Polynesian navigators use dead reckoning, piloting, and celestial navigation, all done in their heads without charts or instruments. With only these, during good weather, experienced navigators and their crews appear at all times, day or night, able to point accurately toward their objective while en route. They also know the directions of home and any reef or island they may be passing even when all landmarks known to them are well over the horizon or hidden in darkness. It seems that on a mental map these well-trained sailors can closely trace their successive positions relative to the known geography virtually without sighting landmarks. Hence dead reckoning in which present position is continually calculated mentally from the course followed and the distance sailed from the starting point is a basic and essential skill. No chart, ruler, or protractor is available to help the traditional navigator's mind and memory.

Conventional piloting is of limited use for Pacific interisland navigation except when near mountainous islands like Truk, Bora Bora, or Hawaii. Observers in canoes usually cannot see low-lying atolls from distances greater than 16 kilometers. Hence in an ocean made up mainly of vast expanses of open water, one is out of sight of land most of the time even on relatively short trips. On the Hokule'a's 1976 voyage from Hawaii to Tahiti for instance, the crew sailed for nearly a month without

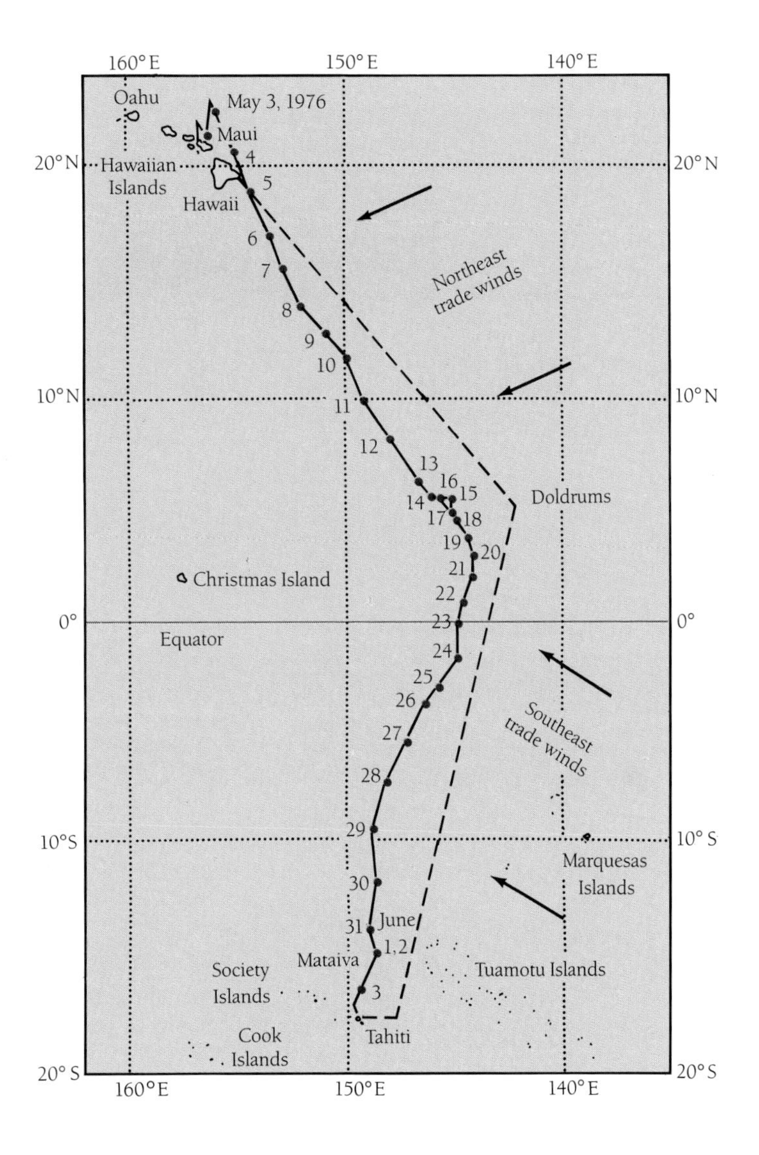

The course sailed by Hokule'a from Hawaii to Tahiti in 1976. Piloted by a Caroline Islands navigator using traditional techniques, the twin-hulled canoe's daily progress is shown by the thirty-one red circles. In the doldrums, a region of weak and variable winds between the northeast and southeast trade winds zones, the canoe made little headway for 5 or 6 days. The dashed line represents the intended course.

sighting *any* landmarks. Helplessly lost native voyagers overtaken by severe storms have drifted for 30 to 40 days without sighting land. On the other hand, piloting *is* surely critical near the beginning and end of a voyage. For instance, an important aspect of setting sail is the use of backsights to line up the right course for a given destination. Part of the traditional pilot's lore is the particular alignments of such local details as a clump of trees, a canoe house, or a point of land, corresponding to

Caroline Islands pilots line up landmarks to set courses when leaving Pulawat for Pikelot, Pulusuk, or Satawal. For instance, setting out for Pikelot, an uninhabited atoll 130 kilometers away, a Pulawatan navigator knows he is on the right west-northwest course when an island tip and the northern edge of two sets of trees appear in line. Three courses to Pulusuk are also shown—two compensate for the effects of westerly and easterly currents.

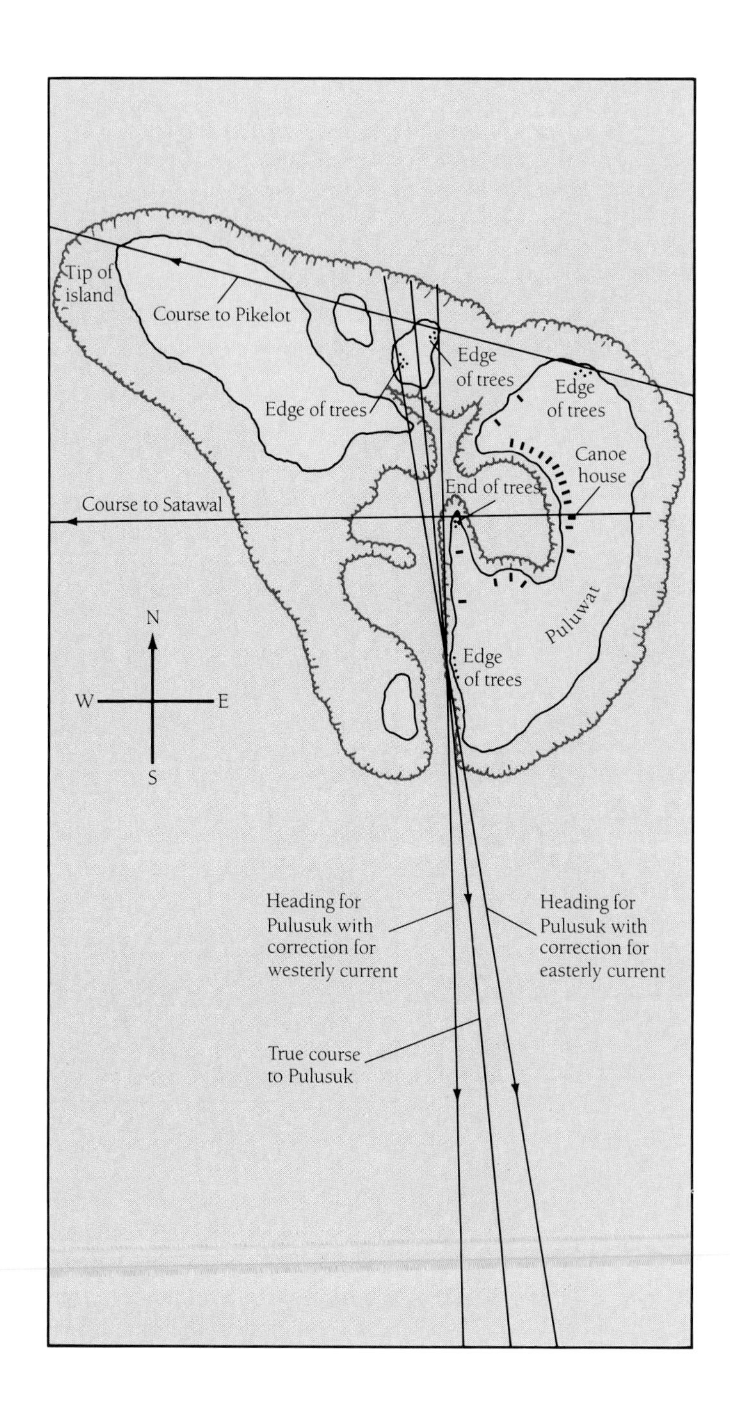

the correct headings toward various destinations. While using such landmarks, the navigators also compensate for drift due to currents, which they sense before starting out or estimate before losing sight of the island. Gauging currents is important but away from land depends on subtle features like wave or whitecap shapes.

The traditional navigators use various sorts of information in steering their courses. Submerged reefs or shoals that can be recognized by water color or bottom visibility may be useful aids for piloting. Knowing about reefs that surround an atoll that is the destination, or about its neighboring islands, may effectively reduce the course precision demanded for a given trip. Awareness of reefs or other land *beyond* the target itself provides further insurance because they may act as a safety net for a wayward canoe that has missed its objective. Traditional navigators often prefer certain courses marked in these ways for travel between two points even if they are longer than alternate routes. Biological clues may also aid interisland navigation. A certain kind of fish or other marine animal may, for instance, be expected near particular shoals or reefs. In places floating leaves or fruits of characteristic land plants may help. Birds such as noddies, boobies, and certain terns that roost on islands at night but feed 30 or so kilometers out to sea during the day are likely to provide information on the bearing of known locations. At dawn and dusk their flight directions will usually be predominantly away from or toward their roosts. However, the behavior or presence of even such tell-tale species may not always be reliable. Many other birds such as petrels or shearwaters commonly seen at sea are no help whatever because they avoid land except for seasonal nesting and breeding. On the other hand, migratory land birds, or shore birds, might give useful directional clues for the experienced navigator in proper season.

A curious and probably biological direction finder used by these Pacific navigators to locate islands has been briefly reported by Lewis. In three different island groups, native pilots have said that oriented underwater flashes of light observed at night far from land at about 2 meters depth point toward islands 40 to 100 kilometers away! Presumably due to bioluminescence, this rather spooky natural compass seems not to have been specifically identified or adequately studied so far. Yet it has certainly been seen and reported as navigationally useful independently in the Santa Cruz Islands, the Gilberts, and Tonga.

Waves and weather as seamarks

Dependable wind or current patterns along a canoe's route can assist in direction finding and in correcting the navigator's estimate of position on his memorized map of the voyage. High clouds over islands are potentially important, too. Such clouds characteristically build up from hot air that rises over land during the day and then move off the leeward side of the island. Sometimes towering 3 to 4 kilometers, these clouds are visible from much greater distances than the low islands themselves. Even

High stationary clouds above islands may greatly enhance a navigator's ability to locate those islands. Because clouds usually trail off to the leeward side and other clouds not directly related to islands are often present, a keen and seasoned navigator's eye is needed to identify the right one.

at night they may be detectable 40 to 50 kilometers away. The knowing and attentive observer may be able to recognize a particular out-of-sight atoll or island from the reflected color of its cloud bottoms or other subtle details.

Wave patterns at sea are one of the most useful navigational clues for the traditional navigator who knows both the seasonal and local geographic features to be expected. At any given time and place on the sea surface several intersecting sets of waves typically converge. Usually these include a predominant swell, originating perhaps hundreds of kilometers away; in many central Pacific areas steady easterly trade winds drive these swells, making them an important reference for direction both day and night. Consequently, a major feature of canoe navigation is the sustained and detailed attention given to the relation between the boat's course and the angle its hull makes to such main wave trains. An important aspect of learning the art of traditional navigation is developing an exquisite deep awareness (some say from the testicles!) of the rhythmic interaction between the vessel and the sea. The slightest alteration alerts the navigator that either his canoe, or something about the wave pattern, has changed. In either case, he must at once make the proper responses. According to Lewis, youthful apprentice navigators in the Marshall Islands are set out at sea in small canoes for first-hand experience with these subtle signs.

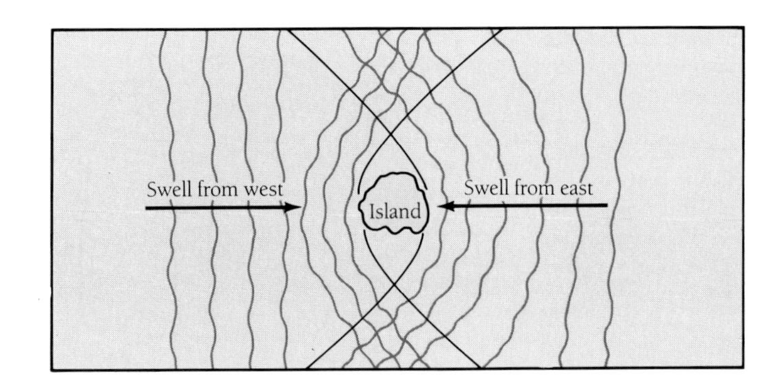

A wave diagram from the Marshall Islands shows two opposed linear swells bending as they approach an island. The curving and interference of the wave trains clearly form a predictable pattern around a single island. When a number of islands are relatively close together and several wave sets are present, however, these relations become quite complex. Knowing such patterns, a native canoe pilot can tell where he is within this system.

An intriguing feature of a long sustained swell is the lens effect islands have on wave direction. As a familiar example, the famous surf at Waikiki usually results from a near 180° bending of the main wave system around Oahu. Consequently, fine surfing waves coming in to the beach typically move against and break directly into the lively trade wind that originally set them in action. More generally this means that islands refract major wave patterns in predictable ways that can be learned. For a successful navigator in regions of the Gilbert Islands or the Marshalls, where there are numerous small islands relatively close together, the local swell directions, as well as the peaks and troughs of possible standing wave patterns, may provide crucial data on a boat's location and heading. Student navigators first learn how to use these clues from the stick charts already mentioned. Thus remembered local wave patterns can function as seamarks for piloting.

Star compass and star path

Using only their naked eyes, the native people of the Pacific islands are good astronomers; their navigators find directions from a star compass and star paths. Within the tropics the apparent movement of stars is considerably simpler than it seems at the intermediate latitudes to which many of us are accustomed. In the northern hemisphere the elevation of the polestar is directly related to latitude. At middle latitudes all stars near the north celestial pole appear to wheel counterclockwise around it while stars far from it rise and set obliquely. Hence both the elevation and bearing of stars seen from the temperate zone appear to change continuously. Essentially similar rotations, risings, and settings are seen in the southern hemisphere but the apparent movement of the stars is clockwise. This rather complex pattern observed at middle latitudes is not particularly easy to follow. Within the tropics, the stars, except near the celestial poles, seem to have a more straightforward course through the sky. They appear at a given point on the eastern horizon, ascend straight up overhead, and then drop down to the opposite location. Thus a star that rises

(*Top left*) Near the equator most stars appear to rise straight up from the horizon to their highest point in the sky then to drop straight down to an opposite point to set. (*Top right*) In the north temperate zone stars appear to rotate counterclockwise around the polestar or follow complex paths from rising to setting. These differences are dramatized by time exposure photos of the night sky in the tropics (*bottom left*) and at the North Pole (*bottom right*).

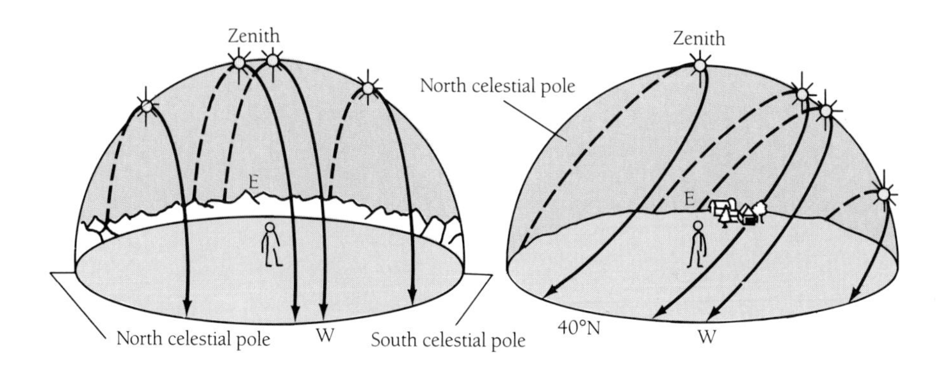

precisely in the east will set exactly in the west after passing directly overhead. In the Central Carolines for instance, Altair, the Big Bird, follows such a pattern. A star that rises exactly in the northeast will set in the northwest, and so on around the horizon.

Not surprisingly then, native Pacific navigators, instead of using wind directions or compass points relative to magnetic north to calibrate bearings, use a *star compass*. Bearings indicated by rising and setting points are named for the corresponding

The star compass that traditional navigators in the Caroline Islands store in their memories indicates no fewer than thirty-two directions. Set down by a westerner, it has the features shown here. As seen from Puluwat, Altair rises in the east, passes directly overhead and sets in the west. Polaris and the Southern Cross, both low in the sky, indicate north and south respectively. The fourteen other stars and constellations combine with these rising and setting bearings to provide a somewhat irregular compass rose.

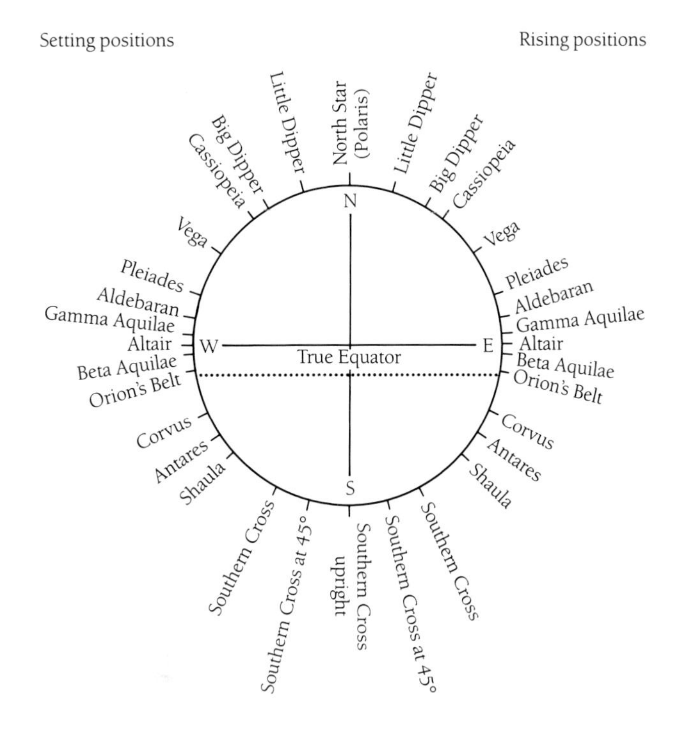

sixteen prominent stars. These stellar compass points are used to calibrate the course direction necessary for travel to the pilot's destination. Typically a canoe pilot may have the locations relative to his home atoll, of fifty or more islands in his mental library. Each of these goals has a known star compass bearing. Native navigators use celestial bodies not only to determine directions but also to hold courses at night. Because pilots can take bearings most easily when the references are near the horizon, they use a sequence of stars rising one after the other at about the same bearing as a sort of linear constellation. By steering from this so-called *star path,* a pilot can hold a constant course all night, a feat that would be impossible if only a single star were used. Ancient Egyptians used comparable star sequences as a nocturnal clock. The naked-eye celestial navigation of the Pacific islanders is rather elegant, but because of its dependence on fair weather, it is still more vulnerable than their piloting. Even moderate cloud cover, which commonly marks at least part of any Pacific day, may severely limit its use. Full cloud cover blocks celestial navigation completely, leaving the canoe pilots to rely on wave and swell orientation and other clues.

To the huge amount of geographical, meteorological, and piloting information already mentioned that our traditional navigator must keep in his head, we must now add the directions of the sixteen primary compass stars plus a sequence of "trailing" stars for each. In addition the stars that reach their zenith over particular places need to be remembered because they can be used to tell whether one is north or south of a known location. The moon and the planets seem to be little used in traditional interisland navigation probably because of their apparent complex movement relative to one another and to the multitude of fixed stars. Accounts of traditional Pacific sailing seem to make little mention of time and its measurement. However, days must be noted en route during extended canoe voyages, and their individual passage gauged by the apparent movement of sun and stars.

The sun may be used during the day in a somewhat similar fashion to a star. However, the solar path, like those of the planets and the moon, is at the same time more complicated and less convenient than those of the stars. Near the equator the sun, like most stars, appears to rise and set almost straight up and down. Yet, being the only celestial body usually visible in the daytime sky, it provides only its own rising and setting directions rather than the full circle of compass points available from star patterns. Unlike the fixed stars, moreover, these solar reference points oscillate seasonally at least 47° between the summer and winter solstices. To use the sun effectively as a compass, proper allowance must be made for such changes. Solar calendars based on seasonal movements of rising and setting points perceived against landmarks are, in fact, common among traditional peoples. However, Micronesian navigators apparently do not employ a solar compass directly but instead calibrate the bearing of sunset with relation to their stable and more versatile star compass. Migratory birds may similarly calibrate *their* compass by remembering the sky polarization pattern at dusk, which depends on the sun's direction as it sets.

No ESP

Recent studies of contemporary Micronesian and Polynesian navigational techniques have turned up no evidence for "mysterious" means of orientation, location sensing, or course keeping. Certainly the skill of the native pilots is close to transcendence, and in the recent past their activities have been associated with sorcery. Nevertheless, no extrasensory or subliminal perception has been evoked to explain their navigation itself. A recent study of British university students might have suggested that the canoe pilots relied on a sensitivity not generally attributed to humans. These subjects, who had no professional training as navigators, were reported to demonstrate an inherent sense of direction attributed to the earth's magnetic field. Although several independent attempts to duplicate the results failed, we will see in Chapter 7 that the possibility that humans can sense the earth's magnetic field is still controversial. On the other hand, extensive experiments on bacteria, mollusks, insects, fish, and birds indicate that animals do sense magnetic fields. Some experts remain skeptical, however, that animals use this sensitivity to navigate.

To equatorial or tropical observers a particular star that reaches its highest point directly overhead is at the same latitude as their location on earth. Altair passes vertically above an observer on Pulawat, for example. Thus a Pulawat navigator at sea knows that if this star reaches its highest point to his north, he must sail north to reach his home latitude. Zenith stars known for various other islands, are also used for this kind of navigational reference. A number of other Pacific star-island pairs are also shown here.

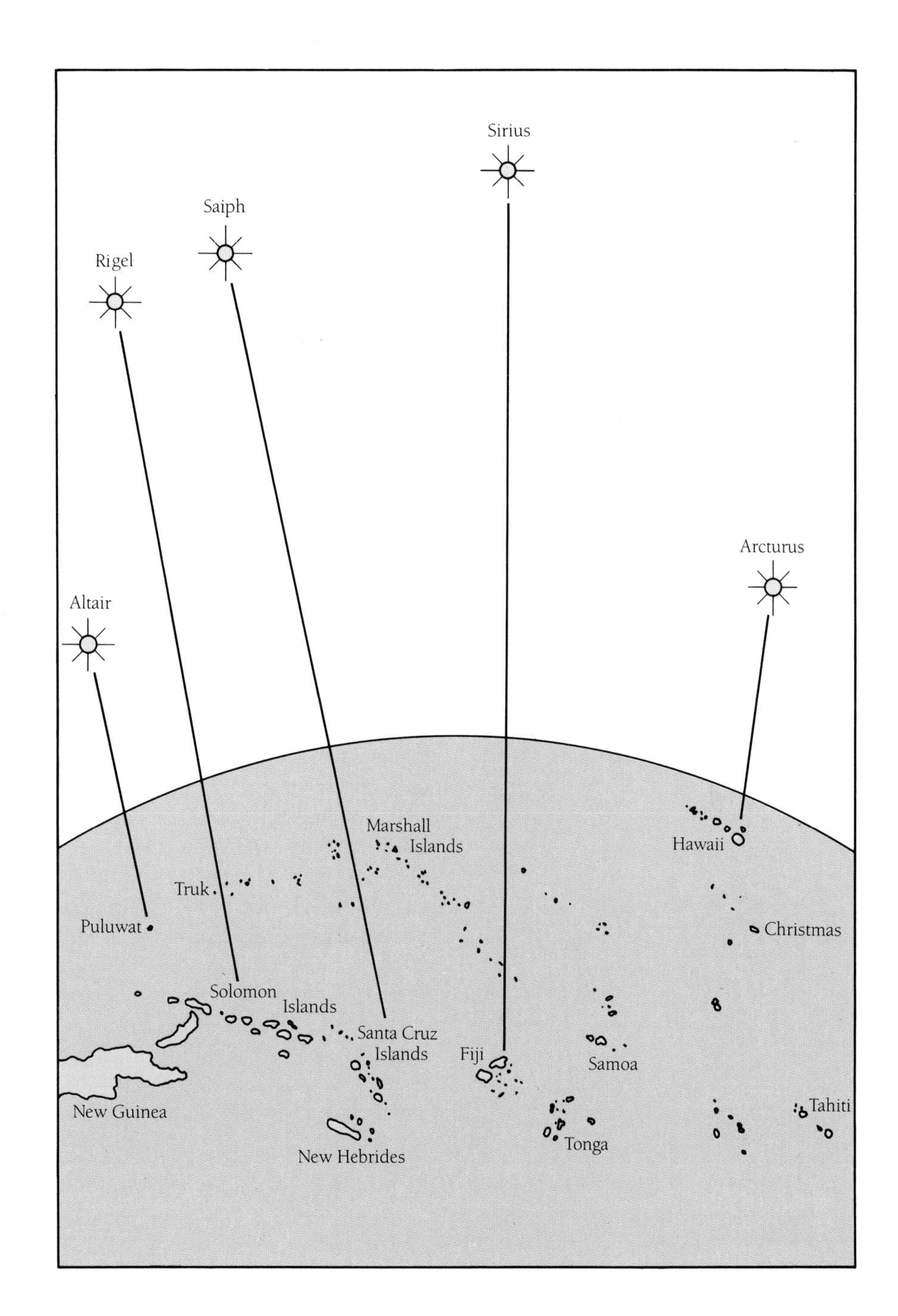

Getting lost

So far we have talked about navigating in clear weather with moderate winds and waves. However, the effects of ferocious winds and canoe-smashing seas must be considered. Practically all the navigational techniques mentioned, especially the important celestial ones, are rendered quite useless by powerful storms. When bad weather prevails or is predicted, sensible native navigators do not begin voyages. They prudently remain on shore. If foul weather overtakes a trip already underway, the crew seeks emergency shelter if a suitable nearby anchorage can be reached. More likely the wind-and-wave-battered canoe will be blown far off course in directions and over distances that can only be roughly gauged by the pilot. Often he will be completely lost, perhaps for a month without landfall.

Not infrequently in historic times typhoons or other violent storms have driven voyaging canoes from the Western Carolines or the Marianas off course by 800 kilometers or so, all the way to the Philippines. In a number of documented cases successful return voyages to the home islands were subsequently made in fair weather, demonstrating the effectiveness of native navigational skills. Differences of opinion still exist among scholars on the importance of accidental or one-way voyages in the dispersal of humans across the Pacific. Indeed some scholars believe that Pacific islanders have been impressive explorers and seafarers not *because* of their navigational skills but *in spite of* them! However, computer modeling has shown that accidental drift seems a highly unlikely mechanism to account for much of the settlement of the central Pacific islands. Detailed analysis of winds and currents makes chance immigration of canoe-sailing Austronesians into eastern Polynesia quite improbable. Furthermore the later spread from central Pacific islands to Hawaii, Easter Island, and New Zealand most likely required deliberate sustained voyaging into the wind. The Hokule'a voyage—a remarkably straight, nearly 5000-kilometer course recently navigated in just over a month from Hawaii to Tahiti by a Caroline Island pilot using traditional techniques to reach a specified small island in a great ocean—has already been mentioned. This trip was repeated in 1980 and 1985, again without instruments for navigation. These records should do much to weaken repeated claims that illiterate native navigators without instruments would be quite incapable of carrying off such a feat. Arguments about details of how the Polynesians discovered the central Pacific islands over a period of 500 to 1000 years may not be definitely settled now. Yet the unquestionable skills of present-day traditional navigators for over-the-horizon sailing speak directly to the point.

Overland trails

The Australian aborigines use linear piloting to travel over their ancestral paths in the interior of the continent. Despite being nomadic hunter-gatherers, these people range widely by following ancient dream tracks or song lines that are central to their cultural heritage and even, they believe, to their individual identity and honor. The native navigator here depends on piloting through a sequence of hills, ravines, hummocks, rocks, and other landmarks. The perceived importance and sacredness of these places comes from their being considered as the spirit dwellings of mythical god-animals such as the rainbow snake, the crocodile, the sea eagle, and the spiny anteater. Little wonder that destruction or disturbance of sacred landmarks by mining, railroads, deforestation, or development has been disastrous to the way of life of these people. As has been suggested for some animal navigation, path following by the aborigines unfolds like a series of snapshots or video scenes known to the traveler. These have been learned and later recalled en route with the aid of heroically long traditional songs whose couplets trace the sequence of directions and landmarks for a given songline. Accordingly, when a journey is made in a motor vehicle, the song itself must be incredibly speeded up from its normal tempo, which is right for walking! Not to acquire and remember this ancient lore signals a grievous lack of character or even soul for these original Australians. Yet few of the recent generation are interested in such primitive mysteries. The navigational skill included is rapidly becoming extinct.

RELEVANCE TO ANIMAL NAVIGATION

Like migrating animals, these traditional human travelers need no instruments or maps to complete their successful long-range travels. Piloting, as well as dead reckoning and celestial navigation for the interisland voyagers, are keenly developed but depend strongly on the oral transfer and individual learning of a large amount of detailed information combined with prolonged practice. The distinctly social and cultural behavior involved in primitive human navigation is rarely to be matched by any animals. Honey bee dances, which communicate directions and distances to hivemates, seem to be quite exceptional among migrators. Although animals, like canoe pilots, do indeed learn from experience, some insects, fish, and birds migrating for the first time do succeed quite well! Does this mean that animals inherit genetic information for navigation while humans do not? Both the Pacific natives and animals in general navigate successfully with no written language and no instrumentation whatever. In both cases the same basic problems have to be solved without aids that seem essential to us. All navigation of whatever kind involves four basic dimensions: places, directions, distances, and times. For us these are embodied in maps, compasses, clocks, and calendars. What are their animal counterparts?

How do the albatrosses do it?

4

Spatial Orientation and Course Keeping

Ordinarily, only a dead fish floats belly up and a fallen sprinter, like an overturned car, can no longer run. One of the most basic and widespread behavioral responses is the *righting reflex*. A disoriented animal restores its reference position in space by this means. Most of an animal's ordinary movements and many of its perceptions depend on its maintaining a standard placement: legs, wings, mechanisms of defense and offense, sense organs and so on, are all localized in certain strategic surfaces or poles of the body. Locomotion, the behavioral element that animal navigation guides, starts from a basic spatial orientation. To begin our analysis of how animals navigate, consider the pervasive effects of position in space and of motion through it. Two universal concepts of classical physics—*gravity* and *inertia*—are needed to understand these functions. Spatial orientation of animals depends strongly on gravity because the attractive pull of the earth everywhere sets up vertical and horizontal reference directions. Because gravity acts on the animal's body and all its various parts, inertia is critical in sensing the direction of this force. Inertia is the property of an object that requires an external force to move it and, if it is already moving, speed it up, slow it down, or deflect it from a straight-line course. Inertia not only acts directly on locomotion but also allows an animal to sense changes in both its linear speed of movement and its rotations about the body axes.

A cat in free fall demonstrates the righting reflex. Even from an upside-down position, the animal twists its body within a fraction of a second to land safely on its feet.

SPATIAL ORIENTATION

If turned over, an earthworm, a crab, a goldfish, or a swallow will immediately try to rotate its body to regain the normal spatial relations with the environment. The need

A bilaterally symmetrical animal has three perpendicular axes, one dorsoventral, another anteroposterior, and the third transverse (right and left). At rest, the dorsal end of the organism's vertical axis points up to the zenith while the ventral end points downward in the direction of gravity to the center of the earth. The other two body axes lie in the same horizontal plane in which compass bearings are defined around a circle. The animal may move in either direction along any of its three axes or may rotate clockwise or counterclockwise around them. Thus these six variables plus time define possible movements for navigation in space.

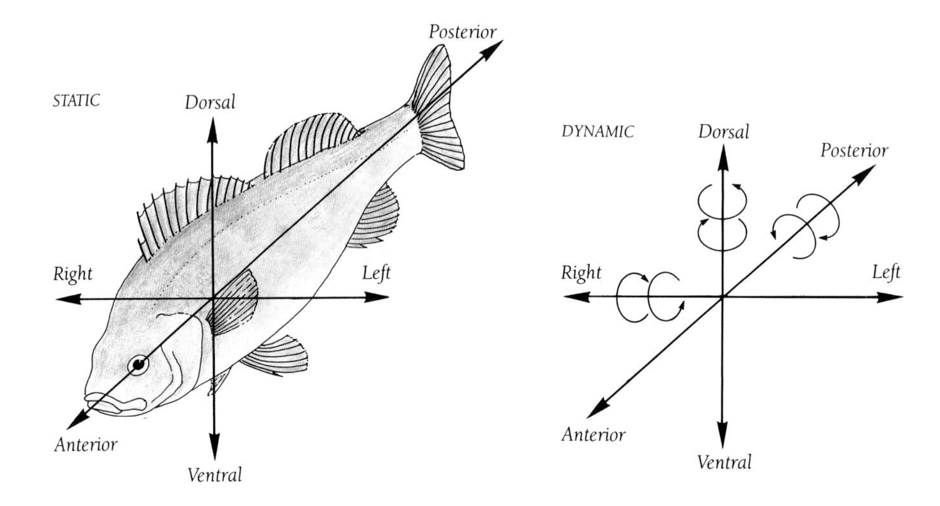

to be oriented in a particular way arises because the bodies of most free-living organisms have a bilateral symmetry defined by three perpendicular axes. These axes mark the anterior (head) and posterior (tail) ends of the body, its right and left sides (usually symmetrical), and its dorsal (back) and lower, ventral (belly) surfaces.

For an animal to function well this body geometry must relate properly to the geometry of the environment. Usually the anteroposterior and left-right axes are horizontal and the dorsal side up. Flying, for instance, is only possible when the wings produce lift and thrust against the air. Lying on its side or back, a bird cannot take off but can only beat its wings ineffectively. Receptors for vision, hearing, and olfaction typically are clustered on the head end of the body. Other special sensors, too, have characteristic body locations. As a result, proper functioning of most perceptual channels needed in navigation also depends on correct spatial orientation. These input channels lead into the central nervous system where external spatial information must be projected onto a sort of neural map. Of course, there are interesting deviations from the typical pattern of basic orientation. Among the bony fishes, for instance, there are upside-down catfish, head standers and tail standers, as well as the whole tribe of flatfish, such as flounder and halibut, that regularly swim and lie on one side of the body so that their dorsoventral axis is oriented horizontally rather than vertically. Humans, as upright bipeds, are also nontypical; we hold our anteroposterior body axis vertically (except for the head) and our dorsoventral axis horizontally. Even for such nonconformists, however, there is a standard body orientation. They, too, have righting reflexes but ones that would not be "right" for most other species.

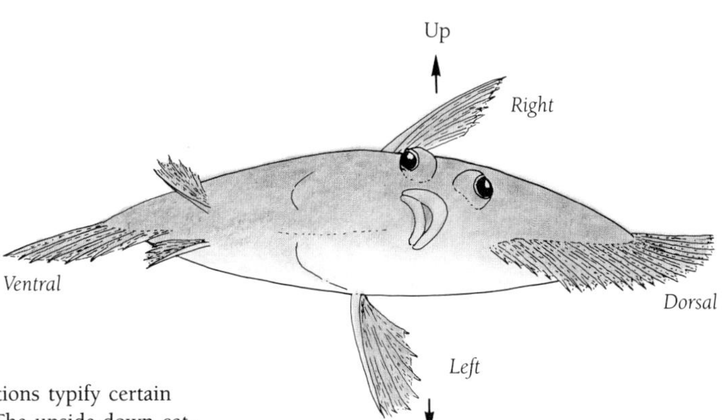

Unusual spatial orientations typify certain bony fishes. (*Top left*) The upside-down catfish *Synodontis nigriventris* is usually positioned with its belly oriented upward. (*Far right*) The leaf fish of New Guinea swims with its head straight down. (*Right*) Flounders and other flatfish maintain the dorsoventral axis horizontally both when swimming and when resting on the bottom. The two sides of the body are quite asymmetrical. One side, which may be either right or left in different species, becomes the top of the fish, boasts two eyes, and changes color for camouflage. The bottom side has no eye and is typically pale or white. Substantial behavioral and structural accommodations, such as the migration of the left eye in the species shown to the right (upper) side of the body, have evolved.

Although it is usually important for animals to orient normally, acrobats and aircraft stunt teams prove that humans with skill and training can maneuver through extraordinary spatial positions. Flies can walk on ceilings and salmon can leap up waterfalls. In fact, all animals may spend much of their time with their orientation in flux as they react to wind or turbulence and carry out particular tasks. Those species that walk or run must also maintain effective contact with the ground whether it is level or tilted, rough or smooth. Their body orientation is thus often a compromise between their standard posture and what the substrate dictates. If the resulting stance is asymmetrical, the legs, head, antennae, or eyes may move to compensate. Usually these mobile parts tend to maintain their normal orientation to gravity despite the skewness of the body axes. There are many examples of terrestrial or

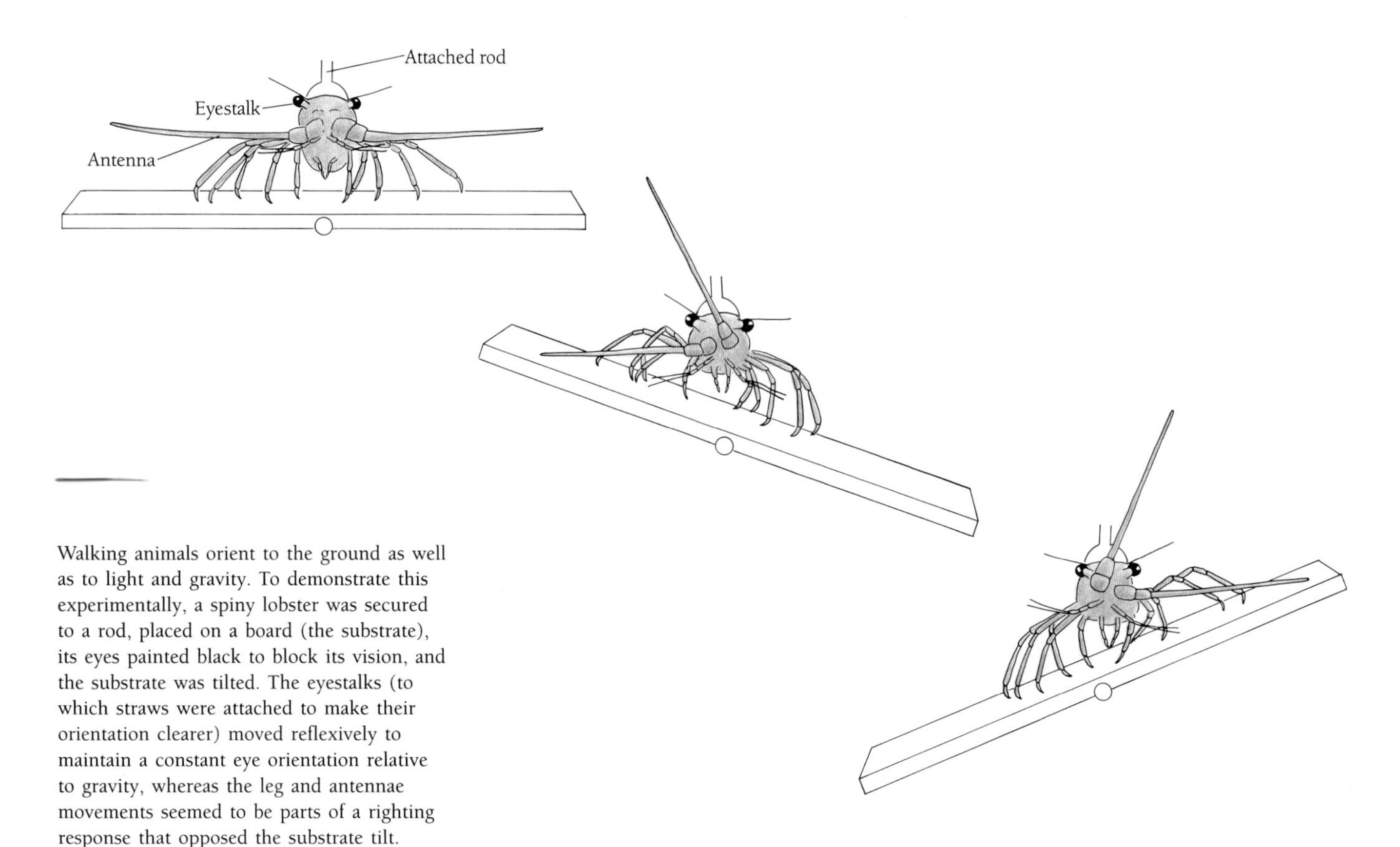

Walking animals orient to the ground as well as to light and gravity. To demonstrate this experimentally, a spiny lobster was secured to a rod, placed on a board (the substrate), its eyes painted black to block its vision, and the substrate was tilted. The eyestalks (to which straws were attached to make their orientation clearer) moved reflexively to maintain a constant eye orientation relative to gravity, whereas the leg and antennae movements seemed to be parts of a righting response that opposed the substrate tilt.

bottom-living species accommodating in this way to conflicting sensory data. However, our main question here is: How do animals generally find and keep their basic axial orientation?

Most of the discussion that follows features animals freely swimming and flying. Because water and air are rather uniform, the righting reflex and spatial stability depend primarily on the direction of gravity, which determines the vertical and the horizontal anywhere on earth. When visibility is poor, as in darkness or in turbid water, reference to gravity may be the only way swimmers and fliers can keep their dorsoventral axis upright. Most animals do, however, supplement reference to gravity with visual aids when they are available.

Visual mechanisms

Orienting to the visual horizon is often important for flying animals when visibility is good, day or night. Insects such as locusts, flies, and dragonflies calibrate their position in space this way. Certain desert and marine birds may do so too because

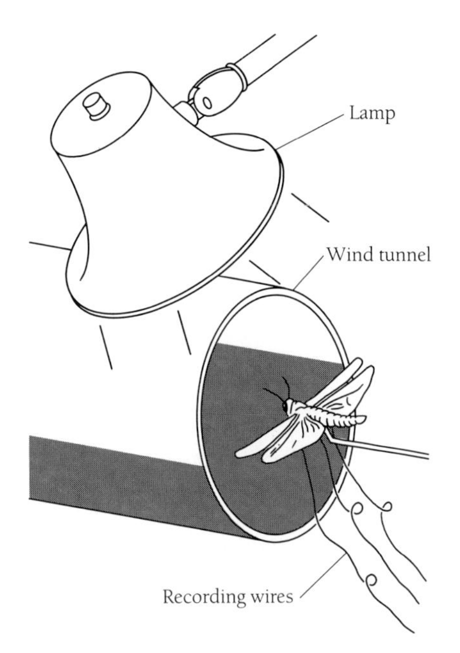

To study whether flying animals orient to the visual horizon, a tethered desert locust wired to recording equipment was made to fly through a wind tunnel. The instrument recorded all rotations of the animal's body. A simulated horizon had been painted on the tunnel's inner wall in a tilted position; when this artificial horizon was illuminated, the insect responded by stabilizing its transverse axis obliquely to match the light-dark boundary and overhead light distribution. In the dark the flying insect did not respond to the horizon and frequently became unstable, rotating slowly around its longitudinal axis.

they have specialized horizontal streaks in their retinas, the receptive layers of their eyes. Semiterrestrial crabs, reptiles, and earthbound mammals also may have such horizon detectors. Even humans depend notably on observing horizontal and vertical edges (such as those of floors and walls) for their sense of space and perspective. A more widespread and better documented visual-orienting mechanism is known as the *dorsal light reaction*. This is a behavioral response to the overall distribution of light and dark in the environment. It was discovered when crustaceans were observed to turn their backs toward the brightest extended area of sky or underwater illumination. The animal reacts then to the whole bright sky rather than to the sun's disk alone. Fliers orient their backs toward the bright sky overhead. Similarly, aquatic animals turn their dorsal side toward underwater light from the sun and sky. Usually the dorsal light reaction reinforces gravity as the animal maintains its basic orientation with the dorsal side up, ventral side down. When the two sensing systems give divergent signals most species compromise. As we shall see, experiments on how the effects of gravity and the dorsal light reaction interact have provided considerable insight into how orientation in general is regulated. But first, the special organs that sense the direction of gravity should be considered.

Gravity receptors

Gravity is a force of attraction acting between two physical bodies. As often experienced it is a continuous pull exerted by the earth on objects on or near its surface. Its effects on orientation and navigation are the same everywhere on earth. The pull of gravity has strength and direction. Its strength, which is commonly measured as weight or buoyancy, depends on the densities of the object and the medium (air, water, or earth) in which it exists and the object's distance from the earth's center. Gravity's fixed *downward* direction serves as a basic reference in animal orientation and navigation.

Specialized receptors have evolved in many different kinds of animals to sense gravity's direction. Typically these sense organs are small bubblelike vesicles lined with cells bearing sensory hairs. Within the cavity lie sand grains, calcareous granules, or tiny bones, all called *statoliths*. Because these are markedly more dense than the fluid that fills the cavity, the earth's gravity acting on them specifically distorts the hairs on which they rest. In many species these statoliths are cemented to the hair tips or embedded in a gelatinous mass that includes the hairs too.

As receptors that transmit impulses to the animal's nervous system, the hair cells react sensitively when they are mechanically stressed. The sensory hairs form a minute, nearly flat cup, or *macula,* on the ventral surface of a hollow, fluid-filled vesicle. Organs such as these that sense the relative direction of gravity from its pull on enclosed dense bodies are called *statocysts*. Sensors of this sort range from units around the umbrella margin of jellyfish to components of the inner ear in all vertebrates including humans. Functionally similar statocysts have evolved quite inde-

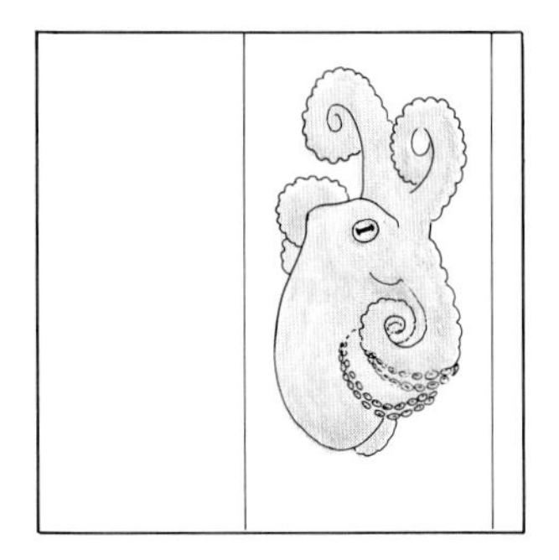
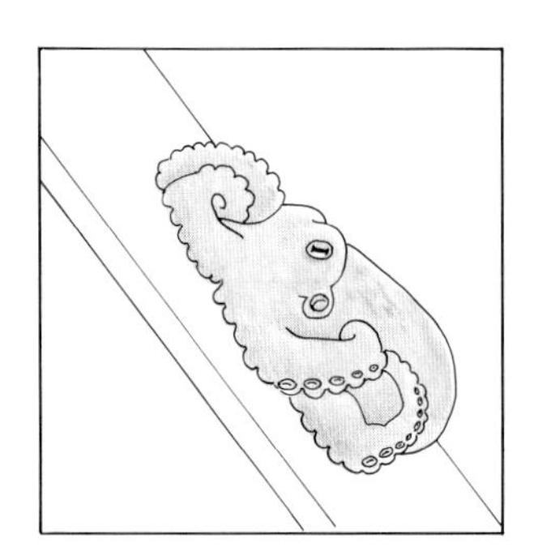

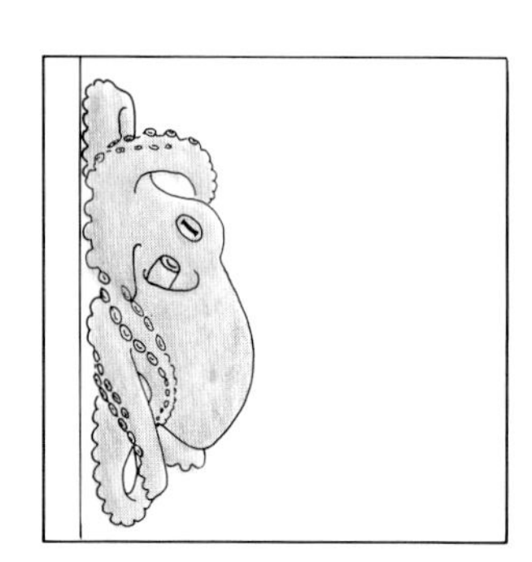

The octopus maintains the horizontal orientation of its slitlike pupil by rotating the eye as the animal changes position. Note that the pupil, and hence the retina, is nearly horizontal whether the octopus is facing upward on a vertical surface (*far left*) or is facing downward clinging to a surface inclined at 45° (*near left*). That this eye rotation depends on gravity receptors is shown by the results of their removal after which the pupil orients to the body's position and no longer to the horizon (*near and far right*).

pendently in mollusks such as squid and octopus, in higher crustaceans such as crabs and lobsters, and in all classes of vertebrates from fish to mammals. These organs typically lie in the head and are paired, right and left. Curiously, they are absent from almost all insects, which manage to control their spatial orientation very well by other means. Receptor units that detect the directional effects of gravity on the insect's head, antennae, abdomen or legs are generally involved. Such movable standard body parts act as stand-ins for statoliths.

Von Holst's experiments

How statocysts function in regulating spatial orientation is well illustrated by some classic experiments on fish by the German comparative physiologist Eric von Holst. Initially he wanted to find out the way gravity acts on vertebrate statocysts. In working on this problem he made some basic discoveries about these organs' contributions to spatial orientation and to the interaction of their input with dorsal light responses mediated by the eyes. When designing his experiments von Holst got a brilliant idea from previous evidence that the direction of illumination can have a strong effect on a fish's position in space. He reasoned that if the gravity reflex and the dorsal light response were made to act in opposition to one another, proper changes in the strength and direction of one stimulus could be used to measure the precise relation between the other system's input and output. This experiment required an ingenious and complicated setup. It allowed the orientation of a slowly swimming fish to be recorded while the direction and intensity of gravity and of illumination were independently varied. Visually alert but tranquil species were needed to serve as experimental animals. Two kinds of freshwater tropical fish, the angel fish *Pterophyllum scalare* and the black tetra *Gymnocorymbus ternetzi*, met the requirements and provided most of the data.

Von Holst's experiments answered his initial question: The force of gravity bends the sensory hairs in contact with the statolith whenever the fish's dorsoventral

Typically both gravity and light control spatial orientation as shown by the drawings of a bony fish. (*Top left*) The downward direction of gravity sensed by the statocysts and the downward direction of light sensed by the eyes reinforce each other, holding the dorsoventral axis vertical. (*Top right*) When the fish is illuminated horizontally from its left, the dorsal light response causes it to rotate about its anteroposterior axis, turning its back toward the light. This tilts the dorsoventral axis in the same direction by an angle dependent on the conflicting inputs from this light response and the response to gravity. Changing the intensity of either the light or gravity will prove this. (*Bottom left*) When the strength of gravity is doubled (but not the light intensity), the tilt of the fish's dorsoventral axis is decreased as the animal's righting reflex receives increasing input from gravity. (*Bottom right*) When only the light intensity is doubled, the fish rolls farther toward its left as the dorsal light response increases.

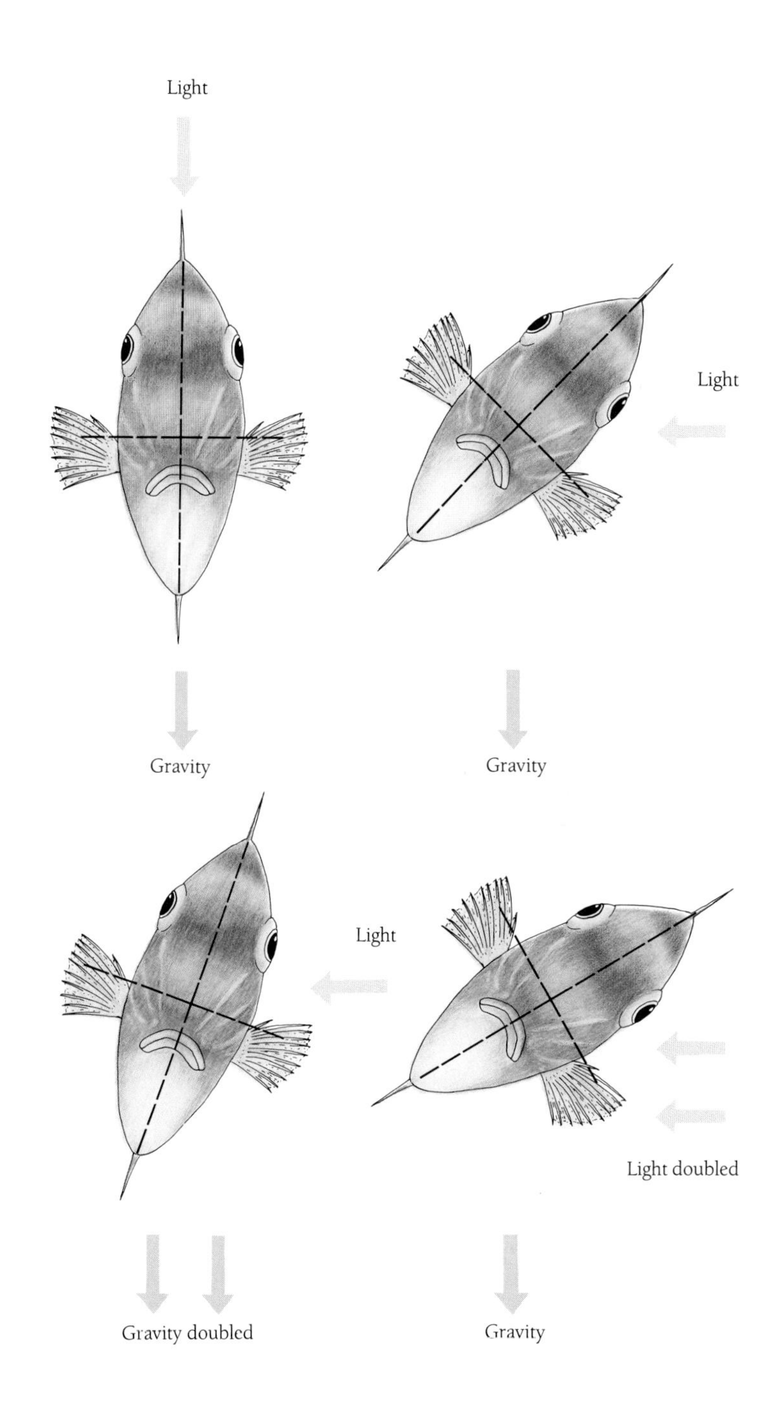

Two hypotheses have been proposed for gravity reception in vertebrates. One notion (incorrect) assumed that the weight of the statolith pressing down on the sensory hairs of the macula provided the stimulus. The alternative hypothesis that shear (or bending) of the sensory hairs is the effective stimulus was proved correct by von Holst's experiments.

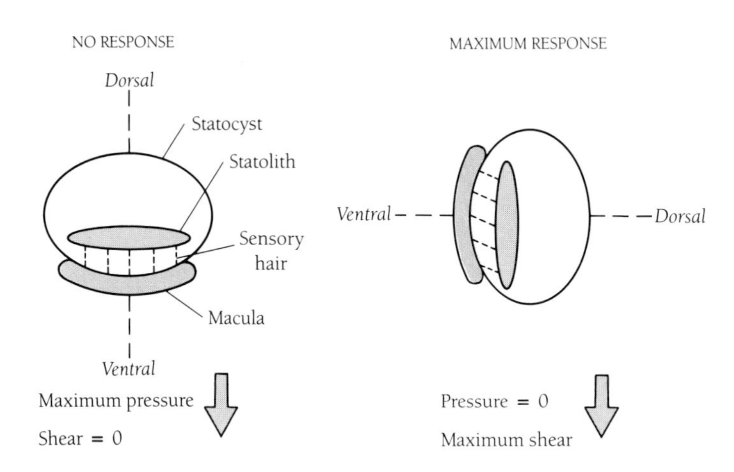

axis does not coincide with the direction of gravity. The greater the tilt of the dorsoventral axis from the vertical the greater the response of the hair cells up to a maximum at 90° when the fish is lying on its side. The bending of the hairs stimulates the fish's central nervous system, provoking the animal to align its dorsoventral axis with the direction of gravity. The structures in the fish's inner ears that serve as statocysts in this response are called *utriculi*. The input to the brain from the right utriculus is just added to the input from the left utriculus. As a result, the strength of the fish's righting reaction is proportional to the sum of the two inputs.

The experiments of von Holst also demonstrated simple stimulus-response relations for the dorsal light response. With constant gravity, the light-dependent tendency to tilt is zero when the downward light direction coincides with the dorsoventral axis of the fish's body. As the light axis is moved from the vertical and as its intensity is increased, the induced tilt increases to a maximum with high-intensity lateral illumination 90° from the dorsoventral axis.

The relative strengths of the dorsal light response and the gravity reaction vary in different fish species. A spectacularly strong dorsal light response is shown by the black tetra, for instance. Upwardly directed illumination alone from a source below the fish can cause this species to turn completely upside down! Its dorsal light orientation strongly overrides its response to gravity if the two are in conflict. In most species the response to gravity prevails, and the maximum tilt that can be induced by light interacting with gravity is considerably less than 90°.

Influence of central state

Further experiments by von Holst showed that the "mood" of an individual animal can strongly influence the interaction between these light and gravity effects. For example, when food extract was added to the experimental tank, the fish became visually more alert as judged from strong dorsal light responses relative to their reactions to gravity. In contrast, when the fish were sleeping in the dark only the gravity reflex was at work. Waking up was marked by the reappearance and progressive increase in the dorsal light response. Such data provide dramatic evidence that even apparently simple behavior may arise from the joint action of two or more kinds of stimulus. In addition, the overall response depends on the state of the central nervous system, which must process the external information received by numerous sense organs. Whether the animal reacts to these signals or not depends on many internal factors including activity rhythms, stage of development, state of arousal, and hormonal or neurosecretory cycles. These diverse and often subtle factors that control behavior may sometimes be directly measured by experiments like these.

COURSE KEEPING

Maintaining spatial orientation while staying in the same place is relatively simple compared with doing so while flying, swimming, or walking. The traveling animal must detect and correct for any deviations that occur in its course. Changes both in movement along a straight line and in angular rotations must be taken into account. As such changes are usually possible along and around any of the three body axes, six variables are involved. A seventh factor, time, must also be included because rates and durations of change must be evaluated too. Hence adequate course control requires the animal navigator to monitor all these possibilities with sense organs, put together the resulting information, and make appropriate responses. Turning around the dorsoventral axis, which determines which way the animal is heading, is central to the next three chapters, which deal specifically with direction finding. Here we need to consider simply how a moving animal maintains a stable straight course—how it maintains a course in a particular direction will be considered later.

Linear acceleration sensors

Statocysts detect the direction of gravity. Other forces that move the statoliths also are detected by statocysts. If, for example, a fish's swimming forward suddenly slows down as it meets a current head-on, the sensory hairs of its utriculi register the resulting forward displacement of the statoliths. In some of his experiments von Holst found that centrifugal force in a fish's environment increases the response to gravity alone. Accordingly in certain situations the central nervous system will be challenged to know whether the speed of movement has changed rather than the orientation to gravity. Although both stimuli can activate a statocyst, animals usually can identify what has happened. For one thing gravity will bend statocyst hairs as

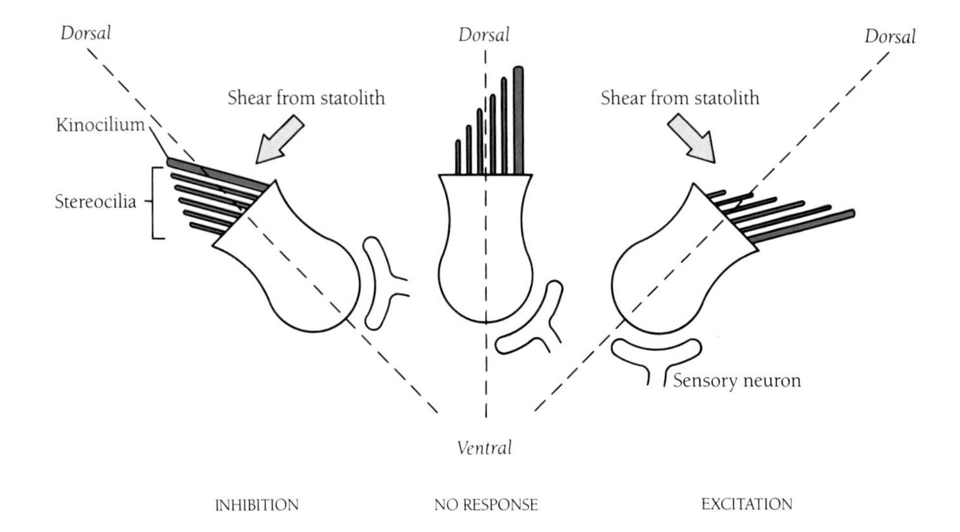

Receptor units such as the macula in the fish utriculus consist of a surface layer of cells each bearing a large number of sensory hairs. One such cell usually has about 30 to 50 shorter, thinner stereocilia and one longer, thicker kinocilium. (*Left*) The receptor unit is inhibited when a shearing force bends its stereocilia away from the kinocilium. (*Middle*) In the absence of shear the sensory cell is in its "resting state." (*Right*) An opposite bend excites the cell. Each of these receptor units signals its level of stimulation via a sensory neuron that conducts impulses into the brain. A resting sensory cell transmits impulses at a certain rate. Transmission of the impulse speeds up as a cell becomes excited and slows down as it is inhibited.

long as the animal remains tipped from its normal upright orientation; the hair displacement due to change in speed will only last as long as the animal's movement is slowing down or speeding up. No stimulus from straight-line motion occurs at constant speed.

Because of their orientation in the head, the utriculi respond only to changes in the rates of horizontal movement. However, two other paired statocysts are vertically oriented in the inner ears of vertebrates so that changes in speed parallel to the dorsoventral axis and those parallel to the right-left axis are also separately sensed. This whole system of receptors thus acts to sense linear acceleration in all three dimensions of space. As cited, two other major animal groups, mollusks and crustaceans, have statocysts rather like those of vertebrates. In all three groups the receptors are activated by shearing forces that bend the sensory hairs. The statocysts of the invertebrates, like those of fish, respond not only to gravity but also to other linear forces. As a rule each set of sensory hairs is strongly sensitive only to motion in a single direction but the whole organ contains an array of differently oriented sets of hairs. Hence a linear force acting on the animal from any direction (up, down, right, left, forward, back) evokes particular nerve impulse patterns in the central nervous system that signal just that direction.

Sensing angular acceleration

Animals in these same three groups have receptors that are specialized to assess rotations—more specifically, they sense changes in the speed of rotation. Most of these sensors are located close to the gravity and linear-acceleration receptors just

Changes in a fish's turning speed are monitored in three semicircular canals of its inner ear. The receptors involved are sensory hairs in a hollow swelling (the ampulla) of each canal. The illustration shows how this works for the anterior canal, which parallels the anteroposterior axis. Change in speed of rotation in this plane moves the fluid that fills the canal (blue arrows). In turn this displaces the hingelike cupula, a gelatinous mass extending across the cavity of the ampula. The resulting shear on the sensory hairs within the cupula excites or inhibits their cells. Thus if the fish's head turns upward, the fluid in this canal flows clockwise, stimulating or inhibiting the receptor cells of the cupula in the anterior ampulla.

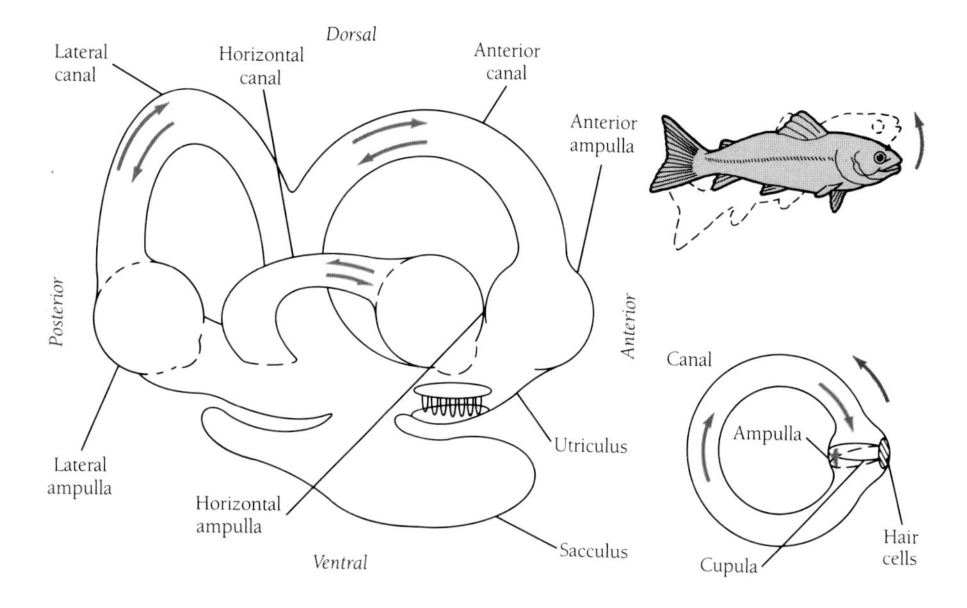

described. In vertebrates they are all in the vestibular system. The rotation detectors are in three pairs of semicircular canals that typically lie parallel to the three main body axes: dorsoventral, anteroposterior, and right-left. When angular speed changes, the fluid contained in one of these curved channels is displaced by inertia. Thus if the head tilts down, the resulting flow of fluid in the anteroposterior canal moves a partitionlike cupula around its basal hinge like a swinging door. This bends the hairs embedded in the cupula, which in turn excites or inhibits the sensory cells of which they are part. The displacement is usually temporary because elasticity soon returns the swinging door to its neutral rest position after rotation stops or reaches a constant speed. This ends the receptor cells' stimulation. Obviously any given angular motion in space must be evaluated by the central nervous system from the information received from all six semicircular canals, three from the right and three from the left sides of the body.

In the statocysts of lobsters and octopuses there are also special rotation-sensing hair cells. Instead of being attached to the statoliths, their sensory hairs float free within the organ's cavity. There fluid flow caused by changes in the animal's angular motion bends these detectors so that corresponding neural signals are sent to the central nervous system. As already cited, insects generally lack statocysts and use other receptors to measure stress on ordinary movable body parts as a means of orienting to gravity. Like statocyst components, these systems are sensitive to changes in both linear and angular velocity. The righting reflex of flying dragonflies

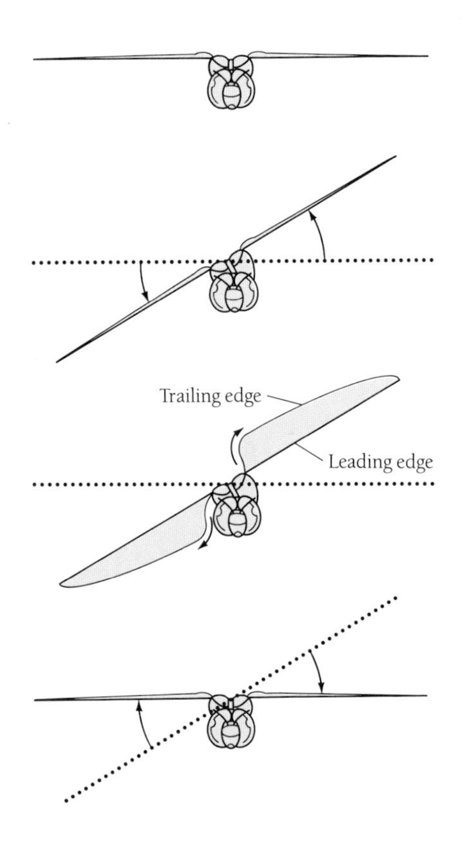

ship between the large heavy head and the slender neck that connects it to the body are key elements of a righting reflex in dragonflies. In normal flight the wings are horizontal and the dorsoventral axes of both the head and the body are vertical. If a dragonfly's wings and body are tipped from their normal orientation, the mass of the head causes it to lag behind so that its dorsoventral axis remains vertical while the body tilts with the wings to the insect's right. This relative displacement of head and neck stimulates hair cells that control wing position. As a result, the insect's wings and body rotate clockwise (as we see it from the front) to restore the normal head-neck relation. (*Top right*) The damselfly *Agrion virgo*, closely related to the dragonfly, has a large head pivoted on a slender neck.

illustrates this principle. Tilting of a dragonfly's body by a gust of wind stimulates receptors in the insect's narrow neck in such a way that adjustment of wing position restores the normal flight attitude.

An insect gyroscope

Automatic pilots of ships and aircraft include a gyroscope, traditionally a heavy wheel rotating at high speeds in a universal ring mount. It has the important property of maintaining its spin axis in a stable fixed position in space as long as it is not subject to an external turning or twisting force. If it is acted upon by such a force, it responds with a displacement that can be used to measure the force's direction and strength. It would not be easy to imagine an animal having access to this useful property were it to require a heavy rapidly spinning wheel as a body part. However, an oscillating structure, such as a rod hinged at one end or even a vibrating tuning fork, develops gyroscopic properties periodically while it is in motion. Such an

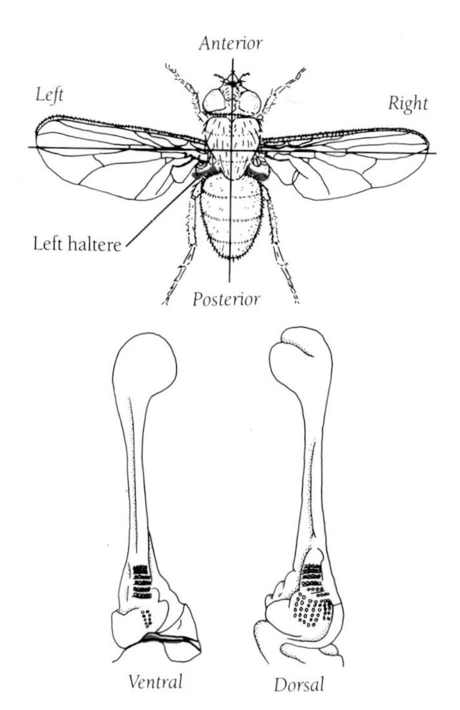

Halteres, which are not functional in lift or thrust, serve as a pair of oscillating gyroscopes in flies, stabilizing their flight around the dorsoventral axis. Receptors (shown as arrays of small circles) in the bases of these vibrating units are periodically stimulated by strains produced when the fly turns at high speed evoking a flight reflex that prevents the spin from getting out of control.

oscillating gyroscope has evolved in dipteran insects in which it helps to maintain their flight direction. When flying, dipterans such as the housefly must maintain stability around the dorsoventral axis even when making extremely fast turns to the right or left. As their name indicates, dipterans differ from most winged insects in having only two wings for flying, the anterior pair of the usual four. The posterior ones are in fact present but are only minute club-shaped elements, the *halteres*. They do not function as wings but are nevertheless critical. If they are inoperative or removed, flies—of certain species at least—tend to be unstable in flight; they may go into a spin and crash.

The wing vestiges oscillate during flight at the same frequency as the forewings and function as oscillating gyroscopes. When a fly's anteroposterior axis, about which the halteres vibrate, is turned to right or left, the gyroscope reacts in response to the twisting force. This reaction stresses a particular region of shell-like external skeleton at the base of these moving halteres and so activates receptors located there. These critical sensors are stimulated only if the fly rotates about its dorsoventral axis while in flight. Evoked by the stimuli, reflex changes in the beating of the forewings restore aerodynamic balance, reducing further rotation. A fly does sometimes make deliberate, very fast turns around its dorsoventral axis when, for example, pursuing another fly. The haltere reflex prevents such rotations from continuing into disastrous spins.

Inertial navigation

Inertial sensing systems, such as the haltere reflex, that are activated by changes in linear or angular velocity are important for animal navigation. Their function, as already mentioned, depends on the basic principle that an external force must act on a body at rest to set it in motion and an external force must act on a body in motion to change its speed or deflect its course from a straight line. An animal's inertial receptors do not require *external* sensory information, but depend solely on internal monitoring of displacements. External forces can be monitored by the movements they cause in body components like statoliths. As we have seen, the receptors involved respond to internal stimuli but not to external ones like light or touch. Navigation requiring no external directional clues (often called *ideothetic* or *kinesthetic*) would seem almost to be a contradiction. Yet there is ample evidence that internal receptors sensing inertia can quite effectively control short-range navigation in certain animals.

The ability of the hunting spider *Cupiennius salei* to relocate a particular place from which it has been chased (as far as 0.75 meter) is a well-studied example of such behavior. Both the direction and range through which it moved are known to the spider without any external directional clues. This behavior is strictly dependent on a special kind of receptor in the external skeleton. The lyriform sense organs

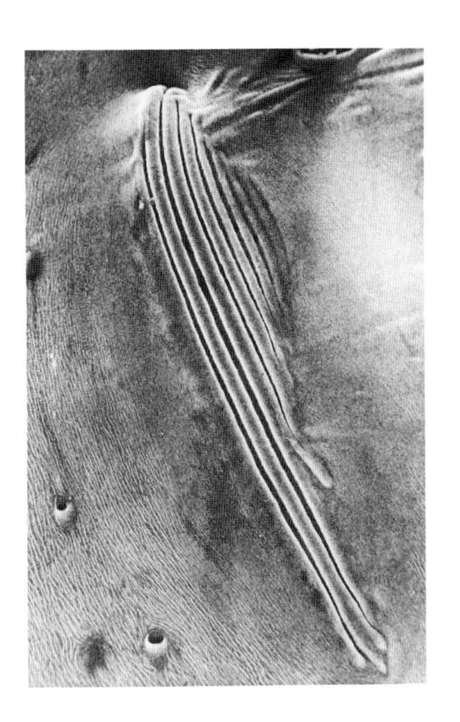

(*Left*) The lyriform organs of the hunting spider *Cupiennius coccineus* magnified 1100 times. (*Right*) These organs, near the joints of the legs, are essential to the hunting spider's navigation.

consist of multiple innervated slits that are near certain joints in the spider's legs. The sensory message of nerve impulses from these receptors somehow specifies the distance and direction the spider has fled, for example, from its abandoned prey. Even though the experimenter has removed the prey, the spider being studied still takes a direct course back to where the food had been, showing that an externally perceived target is not needed to home on. However, this homing fails if some of the lyriform organs are removed from the spider's legs. Just how these sensors of mechanical stress help the animal to relocate its starting point remains to be discovered.

Some animals show a related self-contained ability to resume a given path after *detouring* around obstacles. Honey bees, for instance, can fly around a steep hill instead of following a direct compass course over it. Certain species including birds, if forced to deviate by, say, 45° from a set direction will "remember" that deviation and compensate for it. When they are free to do so, they turn back 45° to the original course. Such behavior with no external compass clues has been documented in the laboratory for myriopods (such as millipedes), in insects, crustaceans, and some birds. Angular acceleration sensors (presumably in the semicircular canals of the inner ears) appear to supply gerbils with data about the complex path followed in moving away from the nest so that they can home accurately without further information.

Such remarkable ability suggests that animals are capable of dead reckoning without needing external sensory input. For the human version of this technique a ship's navigator requires four things: a known starting point, a compass to set the initial course heading and subsequent directions, some means of estimating the ship's speed through the water, and a clock to measure time spent on each leg of the course. The navigator plots the sequence of directions and distances traveled on a chart until reaching the destination. The ability to sum segments of a journey in this way is known in a variety of animals including ants. If the spider and the gerbil can effectively dead reckon using only such *internal* inertial systems as the lyriform organs or the inner ear, they have a navigation technique of great potential value. It could play a major part in foraging and migration, for instance. Given that two such remotely related organisms have this capacity, how many others share it? If adding directions and distances works well for terrestrial animals (such as spiders) over short ranges and short time intervals, *how far* and *how long* can it be extended to serve global navigation?

The possible importance of inertial navigation in animal migration has, of course, been repeatedly considered. As we have seen, vertebrates, crustaceans and mollusks do have sense organs for measuring changes in linear velocity along, and in rotation around, their three main axes. They also have central nervous systems to evaluate all possible accelerations of their body. From such data both the *distance* traveled and the *direction* made good can be rather easily calculated. Hence an inertial system would seem in principle to provide the information needed for dead reckoning. A number of practical problems, however, may limit its effectiveness in animal navigation over more than short ranges and durations. For example, detection of very slight accelerations would be needed as would a keen ability to discriminate both random disturbances (such as turbulence in wind or water) and the effects of gravity from the speed changes important for inertial navigation. Without such discrimination winds and currents, for example, would introduce serious errors for fliers and swimmers. Over long distances any errors arising in the system would pose the threat of being cumulative and growing intolerably large over time. On this account skeptical biologists have been hesitant to accept inertial navigation as important for long-range animal movements and migrations. Nevertheless, in human engineering applications, say in a nuclear submarine, such systems do operate over long periods like weeks or months with astonishing accuracy. Also, it is true that modest continual corrections of an inertial mechanism could be made by animals, for instance with some kind of direction finder such as the bearing of sunset or landmarks. These would minimize cumulative errors and greatly increase the system's usefulness in global navigation.

So far, we have seen that sensory receptors, including eyes in the dorsal light response, help animals in both the static and dynamic aspects of controlling spatial orientation. Now the *dynamic* effects of *vision* on orientation need to be considered. These complement the movement-analyzing functions of statocysts and other such receptors. Like the dorsal light reflex, they provide an optical way of knowing what is going on. Visual detectors of changes in speed respond to the movement of images on the retina. The behavioral response to moving stimuli depends among other things on image size as well as the direction and pattern of displacement. Such visual data can help to stabilize flying and swimming; they can also improve navigation otherwise dependent on inertial sense organs, especially in the face of winds and currents.

Optomotor responses

Almost universally, eyed animals turn when a *large* area of their visual field moves. In this *optomotor reaction* the body, head, eyes, or eyestalks rotate in the direction of motion as if to fix the large image on their retinas. Such responses are quite distinct from tracking behavior in which the animal's gaze follows a *small* moving targetlike object, say a bird, crossing an extended background area like the sky. In that case the object appears to be stabilized in the visual field and the background seems to flow by. Between these extremes, *moderately large* images in motion often evoke still another reaction: escape or collision avoidance reflexes. We ourselves commonly show optomotor responses in eye-movement patterns when we watch scenery pass by a car or train window. Then a sequence of eye rotations occurs at a rate that periodically tends to fix the retinal image. When the comfortable limit of eyeball movement is reached, the eyes reflexly snap back to the initial line of sight and again follow the apparent displacement of the large area seen. From a strictly visual point of view the observer does not know whether the car is driving past the scenery or the landscape itself is rushing past the car. The laboratory equivalent of this sort of response to perceived motion is usually evoked when the observer or experimental animal is surrounded by a rotating cylinder with vertical black and white stripes. In this situation a flying insect, a swimming fish, or a walking crab or mouse all tend to turn to follow a pattern displacement. When the body is restrained, the head, eyes, or eyestalk, if movable, will follow the stripes with a comparable periodic recovery.

The navigational significance of the optomotor response is that it helps to keep the animal on a steady course of movement. Typically, all or most of the objects in its visual field would appear to move only when the animal is accidentally deflected, as by turbulence or when it voluntarily changes its heading, say from east to southeast. The distinction between these two cases (accidental and deliberate) is clearly of great importance although how the organism actually discriminates them remains largely unknown. Suffice it to say that when an animal voluntarily changes course, the optomotor response is somehow canceled so that the previous heading is not restored. In contrast when buffeting by wind causes gross displacement of most objects in an animal's visual field, it characteristically reacts by turning to follow them as already described. Such responses tend to restore the animal to its previously set movement direction.

Responses to displacement by current or wind

If aquatic animals are being carried sideways downstream they turn to face into the current and then swim against it, maintaining a fixed position over the bottom. Similarly in air, flyers turn into the wind. This behavior may be important in preventing the animals from being washed or blown away. Such responses to current and wind depend primarily on visual mechanisms; whether other types of sensors are also involved is unknown. When animals are navigating rather than simply staying in one place, the situation is more complex. Optomotor responses, reactions to current or wind, and course keeping are all intimately related. In nature a flying or swimming animal moving ahead sees an actually fixed array of visible objects apparently flowing toward it, passing it, and then falling behind. Because of its own progress, such apparent movement of the environment appears as a symmetrical passage across both retinas. This does not evoke an optomotor response unless the rate of movement on the two sides is not symmetrical or, in the extreme, is reversed. When accidental or forced turning causes that to happen, an optomotor reaction will tend to put the animal back on its original heading. Note that turning and moving ahead evoke different apparent motions. For rotations the whole visual field appears to be displaced at the same rate. For linear movememt nearby objects appear to move faster than more distant ones.

If winds or water currents parallel to an animal's course move at rates comparable to speeds at which the animal is moving, the resultant change in speed over the earth will be visibly obvious provided there are landmarks to be seen or in some cases waves on the sea surface or sand ripples on the bottom. If flow of the medium is faster than the animals' flying or swimming speed, the apparent sequence of landmarks will be from back to front. This induces a locomotor adjustment in which increased locomotor speed matches or reverses the downstream drift perceived. Actually during migrations in good visibility, birds tend to increase their flying speed in a head wind and reduce it with a tail wind. This means that they must visually stabilize their rate of motion over the ground. For insects, too, both wind-tunnel experiments and field data show that a number of species use optomotor data to hold a constant ground speed.

The apparent movement of landmarks is also important for course keeping and goal reaching by animals moving in winds or currents that are not parallel to their route. The actual travel achieved is the sum of the animal's own movement through the medium plus the added effects of the speed and direction of crosswinds or crosscurrents. The resulting drift can have a disastrous effect on navigation, unless it is taken into account. Many sailing ships were lost because up to the end of the eighteenth century navigators relied on dead reckoning without being able to correct for drift. As we have seen, animals may use optomotor responses to improve their course holding; some terrestrial and flying species can also use receptors located on their heads, particularly in hair patches on antennae, to detect wind direction, ad-

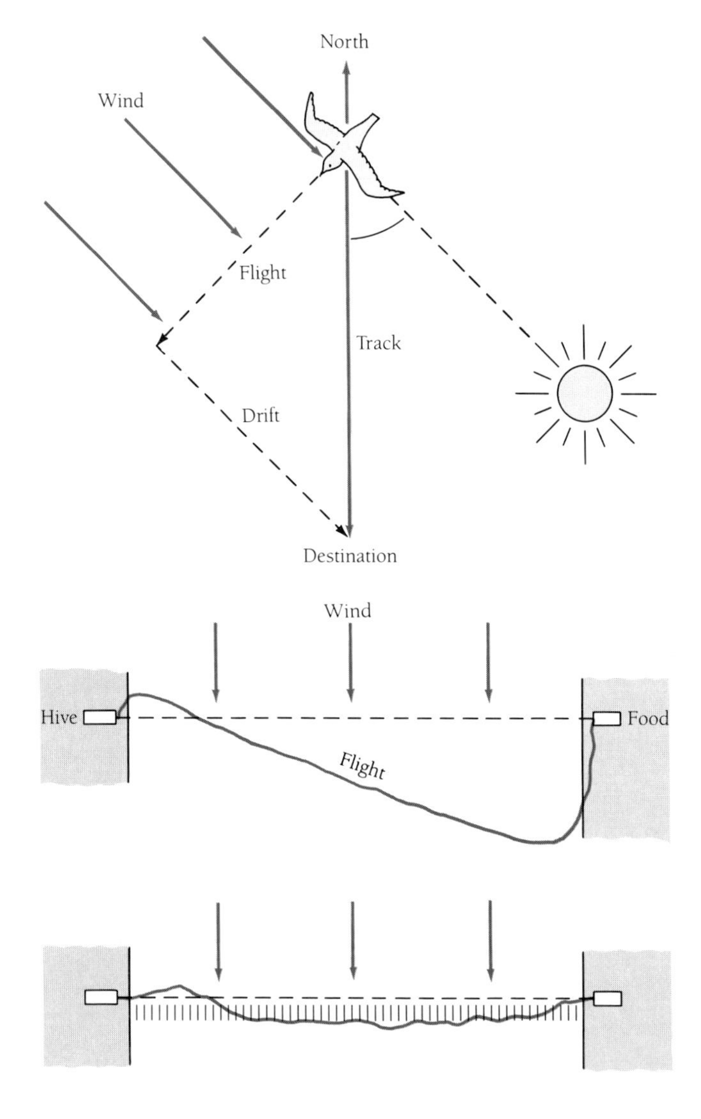

Flyers and swimmers must be able to hold a course against a strong wind or current. (*Top*) For example, suppose a bird must reach a destination that is 45° to the right of the sun. A strong wind toward the sun would carry the bird far off its course if it heads directly for its destination. To stay on the track, the bird must gauge the effect of the wind by observing the apparent drift of the landmarks beneath it. In this case the bird might estimate that a course 45° to the right of its goal would be most effective. (*Bottom*) Honey bees demonstrate a simpler way of course holding in the wind. The bees were trained on windless days to feed on the other side of a lake that offered them no landmarks. When they had to travel against a wind, the bees flew obliquely across the water and had to locate the feeding place after they reached the other side. But when the shore on which the hive was located was connected to the side that had the feeder by a straight marker made of floating slats, the insects' errors due to the wind decreased because they consistently corrected from this landmark.

justing their orientation accordingly. Honey bees are able to compensate for drift visually by using an extended landmark to correct their flight tracks.

Optomotor responses, current and wind reactions, and drift monitoring all depend on the visual perception of movement. Eyes are quite capable of perceiving movement but only if an image is formed on the retina and two or more neighboring sensory untis are stimulated in sequence. In camera eyes, which characterize vertebrates and cephalopod mollusks such as the squid, the image, or some part of it, must move the distance separating at least two of the receptor cells that line the back of the eyeball. In compound eyes, such as crustaceans and insects have, the image must move at least from one facet on the eye surface to the next. In both cases a network of nerve cells detects movement and analyzes its direction. For optomotor and other responses that require an apparent flow of objects sensed over a large part of the total visual field many such elementary movement detectors must be coupled together to evoke the characteristic turning behavior. In flies, such as the housefly, in which this motion-processing system has been elegantly studied, easily recognizable giant neurons do the large-scale summing up of the sensory input. These same large nerve cells are essential for visual fixation and tracking of small targets as well as for perceiving hard-to-see objects against a confusing background.

Spatial orientation and course keeping may for several reasons seem a somewhat elusive introduction to how animals navigate. First, except for gymnasts, high divers, or ballet dancers our body stance and balance are reflexively taken care of with little deliberate awareness. The pervasive importance of spatial perception for human behavior only becomes obvious if the semicircular canals, or other parts of the inner ears, malfunction. Severe physical disability ensues as it does also when gravitational, inertial, and visual stimuli become scrambled for passengers aboard a rolling, pitching ship. Second, most people do not have much feeling for dead reckoning without instruments beyond youthful experience at blindman's buff. Explorers and adventurers who have tried to walk a straight line with their eyes covered or in a dense fog report little success. Yet the sensory means for such course keeping would seem to be potentially present in many animals as well as humans. A third deterrent to easy familiarity with spatial orientation and course keeping is the need to use the concepts of gravity, inertia, and optomotor responses in explaining how the behavior is stabilized. Even contemporary physicists, primarily concerned as they are with high-energy particles or solid-state electronics, may have to think twice about the kind of Newtonian mechanics involved. This does actually provide a good explanation of the biophysical events in question but also exacts some attention to detail.

5

The Compass and Visual Direction Finding

Which way? Choosing the right direction to go is vital to an animal's navigation. Sometimes a choice may be rather simple—particularly if the goal can be readily perceived. At other times the choice is challenging and complex, as for a long-range migration. In either case the proper heading of the animal's locomotor axis is the key variable. The orientation, at its simplest, is directly *toward* an objective such as a mate, food, or shelter, that is the navigational goal. Remember that "objective" and "goal" here imply merely that the animal's physiology and behavior are under the control of some setting, like a thermostat. The "goal" of a thermostat is the temperature we select, which may be permanently fixed or adjustable. In this sense the animal's goal is the setting, selected by evolution or learning, of its homing or migratory controls. To establish the proper heading toward its goal, an animal needs a compass. With a compass the locomotor axis can be correctly oriented in the environment. The turning and steering responses concerned are stereotyped reflexes that couple sensory information about directions with the control of locomotion. Note that orientation *toward* an objective is not the only possibility. In other circumstances relations to the goal may be reversed so that the animal moves *away* from home or from a predator or other localized danger. These two possibilities can be thought of as positive and negative forms of a basic response. More subtly, as in direction finding from the sun, a course may be steered not just toward or away from a reference but at any given angle relative to it. This third alternative is the most flexible and resembles our use of the magnetic compass to navigate a ship or aircraft. From the fixed geographic reference any course around the horizon, not just magnetic north, may be accurately steered. The reference bearing itself is not the goal in such a case but serves as a navigational aid to finding the goal. The difference

Dragonflies, such as this *Sympetrum striolatum* seen here head on, are visual predators that forage on regular patrolling flights or wait to take off when prey appears.

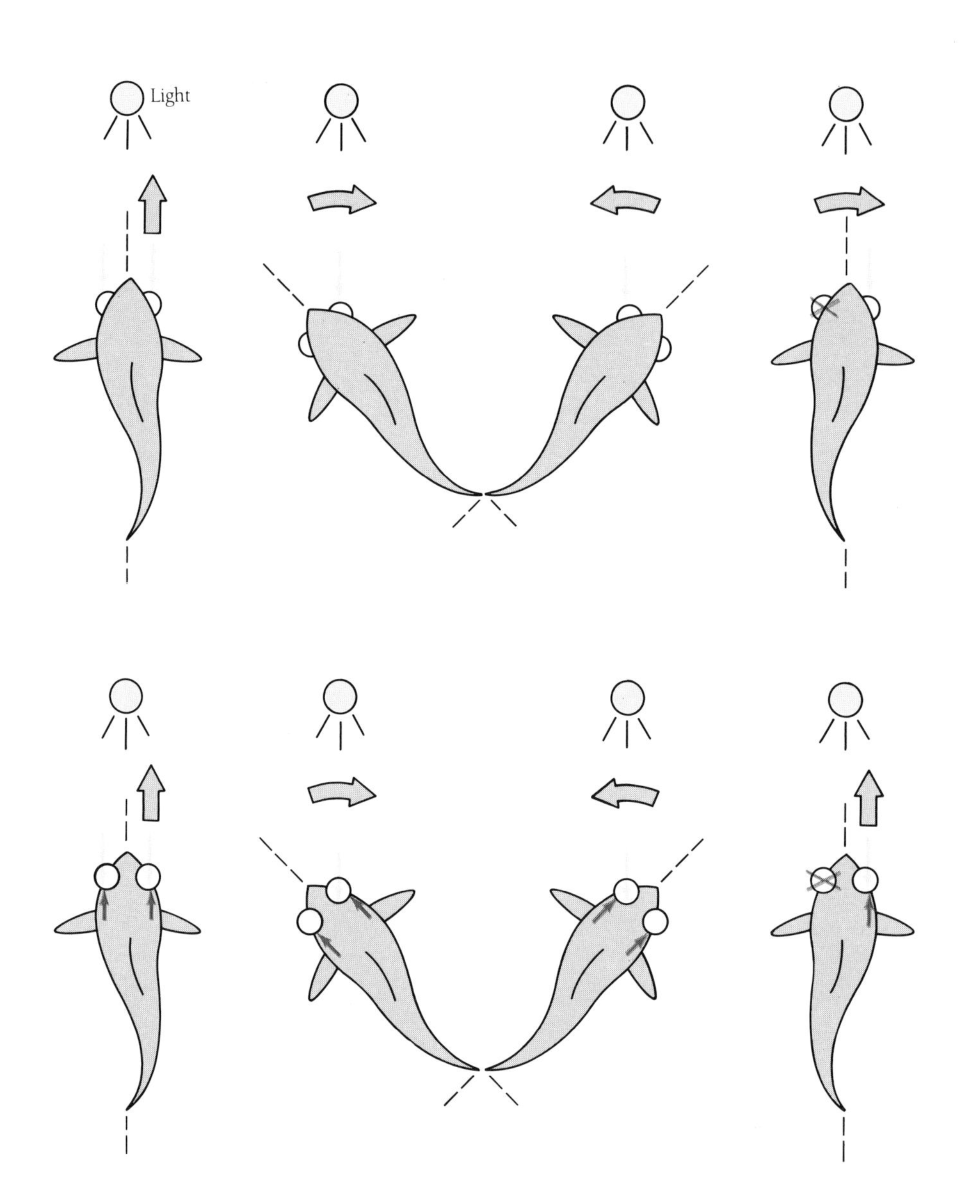

How animals use a localized sensory stimulus to maintain a course can be illustrated by two ways a fish uses a light source to steer itself. One way depends on the fish turning its locomotor axis to equalize the intensity of the light reaching its eyes. (*Top far left*) When the fish's anteroposterior axis is aligned with the light source, the intensity of light reaching its eyes is equal, and the fish swims straight ahead. (*Top near left and near right*) If the illumination of its eyes is unequal, the fish turns in the direction shown by the broad arrows to equalize the illumination. (*Top far right*) If, for example, the fish's left eye is covered, it circles continuously. A second orienting mechanism depends on steering to maintain the stimulus (light source) in a particular place on the sensory receptor (the retina). (*Bottom far left*) When the direction of the light coincides with the reference point on the retina, the fish stays on course. (*Botton near left and near right*) If the stimulus image is not in the reference location, the fish turns to make it so. (*Bottom far right*) Because the perceived location rather than the intensity of the stimulus is crucial, the fish can stay on course even if only one eye is functional. This enables the fish to steer a course that is either oblique or parallel to its locomotor axis, hence it provides a versatile sensory compass.

between steering by the goal itself and using an independent directional reference may seem subtle in some cases. Yet here the terms animal compass and compass response refer specifically to such mediated direction finding. To set its course in this way, the animal must know and measure the angle between the reference direction and the goal's direction.

DIRECTION FINDING

What happens when an animal begins the simple positive response of walking, flying, or swimming straight toward its home? The objective must first be detected and identified whatever sense (vision, smell, or hearing, for example) is used to perceive it. The animal may recognize home by some characteristic signal, such as a color pattern, odor, or the calling of a mate; as stimuli, these signals can act in two ways. One is a *kinetic* effect in which the animal is aroused by the stimulus, moving faster, turning more often, or becoming generally more attentive without showing any specific orientation. Or the stimulus may have a *directional* effect that is more obviously related to our interest here in animal compasses. This mode requires signals that can be perceived from a distance with sense organs able to localize their source. Our image-forming eye is a good example of such a directional receptor. For instance, it shows at a glance that a given object is directly ahead or far off-course to the right. But before considering this basic type of direction finding in more detail, some less obvious course-controlling behavior should be described. This involves responses based on pervasive stimulus gradients, but not on sensing a localized objective from afar.

Orientation to gradients

If free-moving animals have access to an environment with temperatures extending continuously from hot to cold, they tend to stay within a narrow thermal range that they prefer. The actual temperature chosen varies greatly between species that live, for instance, on a tropical coral reef and those in polar seas. Animals also choose environments for particular levels of humidity, light intensity, the color spectrum, or salinity. Thus a strictly freshwater perch behaves as if it prefers negligible salt in the water around it, whereas a brine shrimp from the Great Salt Lake selects water nearly saturated with salt. Such choosing often depends simply on a kinetic response to a continuous gradient in which the speed of randomly oriented travel varies directly with the difference between the actual salinity, temperature, or whatever the animal senses and the level it prefers. The closer the environment to its particular ideal, the more the animal slows down, making it more likely to stay where it is rather than being carried away by further random movements. Hence the gradient of temperature or salinity, for example, provides by its orientation a course-finding signal. Crude as they may seem for navigation, such responses within a limited environment will result in animals locating and remaining in the particular area that has the right adaptive features and their moving away from nonadaptive or hostile places.

An intriguing case of navigation by chemical gradients is shown by the motile bacterium *Escherichia coli*. Although relatively simple in structure, this single-celled microorganism can locate nutrient substances and avoid toxic ones with behavior that is elementary but remarkably effective. Aggregation and dispersal are managed through the interplay of two modes of locomotion. Propelled by the vibration of hairlike flagella, *E. coli* can swim either quickly (about ten body lengths per second) along straight lines or with irregular rapid rotation called *tumbling*. Ordinarily several seconds of linear progress alternate with a fraction of a second of wild tumbling. Because of this sort of frequent shaking of the directional dice, the bacteria move about in randomly directed zigzags and do not get anywhere. However the situation changes when they detect chemical gradients in the surroundings. As the concentration of some substance "attractive" to a bacterium increases, the frequency of its tumbling episodes decreases. Hence each organism will tend to move toward the center of a favorable region. If one overshoots this optimal area by moving into a region having a lower concentration of the substance, its tumbling periods become more frequent, a pattern that eventually will return the organism toward the center again. The responses to a repellent or toxic substance are just reversed. Tumbling increases as the bacteria swim into higher concentrations and decreases as they encounter lower amounts. They thus avoid unfavorable areas and collect in favorable ones. Changes in the frequencies of the two types of swimming behavior turn random movements into responses that seem like attraction and avoidance. Many animals using various senses respond positively and negatively to specific stimulus gradients. We will return to this later in some detail when chemical direction finding is considered more generally.

Visual targets—fixed and moving

A more familiar kind of direction finding depends on visual targets. What we need to consider here is how an optically orienting animal aligns itself to travel toward a seen object whether that goal is *fixed* or *moving* in space. Behavior for homing on a fixed goal might be expected to be different from that for tracking a moving target, but the basic mechanisms involved seem to be closely similar in certain animals at least. A free-flying housefly *Musca domestica* will turn and fly toward any small isolated dark object seen against a uniform light background. Experiments have shown that a fly rotates around its dorsoventral axis to face such a fixed target, thus aligning its anteroposterior axis toward the goal. To pursue a moving target the insect may have to change course continually as when one fly chases another over a dazzlingly fast, convoluted course. Usually such following behavior involves males chasing females with copulation as the goal. Steering by the pursuer is dependent rather simply on how far its heading differs from the goal's direction and whether that course error is changing.

More complex chasing behavior has been found in other fly species. For instance, pursuit behavior by the male hoverfly *Syritta pipiens* appears more versatile than that of most flies. This species can hold its flight location in space fixed for some minutes; it can also, like a helicopter, fly in all directions independent of the way it is facing. This obviously requires exceptional flying skill as well as elaborate

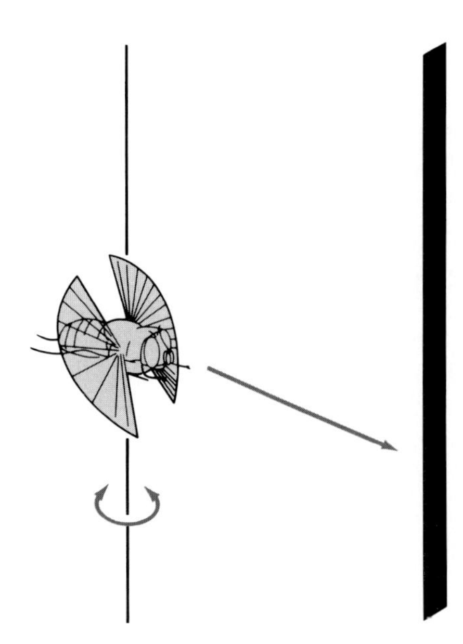

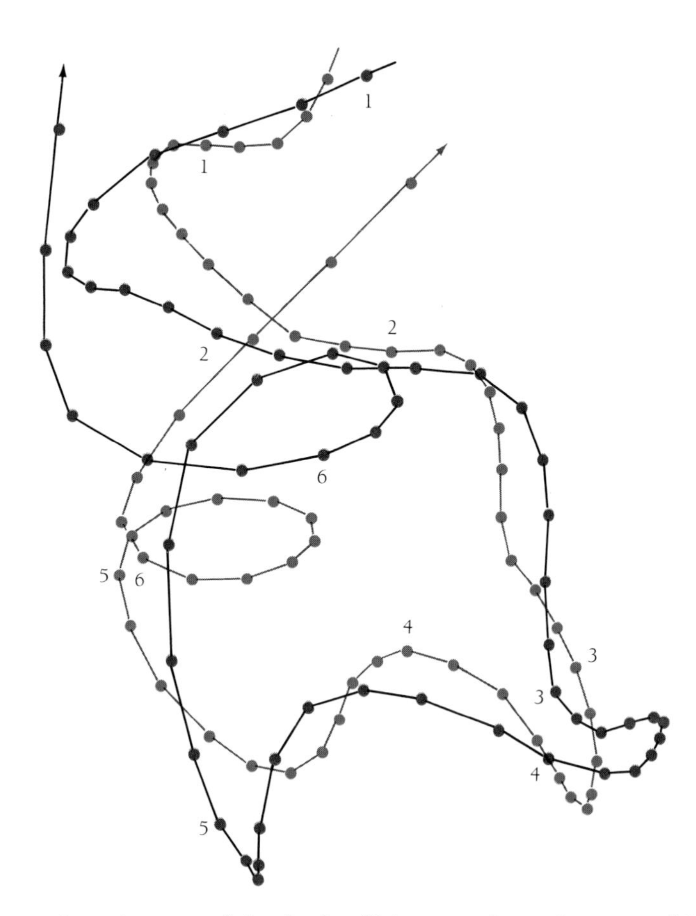

(*Left*) A housefly *Musca domestica* spontaneously steers toward a dark object such as a vertical black stripe in an otherwise featureless background. (*Right*) The flight path of one housefly (line connecting red circles) pursuing another (line connecting blue circles). The chase begins when the pursuer is at point 1 and the pursued fly is some distance ahead, at its own point 1. Successive circles are separated by 0.1 second, and the seconds are numbered 1 through 6.

orientation control. In chasing flights smooth continuous tracking of a target occurs only when its image falls within a visual field of the pursuer looking somewhat downward and not more than 5 to 10° in horizontal extent. When something is seen outside this small region of the visual field, the fly rapidly turns about its dorsoventral axis to bring the target's image within the retinal area specialized for pursuit tracking. During cruising flights without any target in sight, the fly sometimes rotates quickly as if searching with this area, which scans over wide angles as the animal turns. The alternation of smooth tracking and jumpy turns is quite similar to human eye movements. A small central retinal area, the fovea, plays a key role for our detailed bright light vision, too. We rapidly rotate our eyes as well as turning our head and body more slowly to bring images into this area. Like the hoverfly pursuing a moving target, *we* have two ways of visual fixation. Objects in motion already seen in the fovea are followed by a slow smooth tracking response. In contrast, objects sighted by the peripheral retina are brought to the central retina by very fast stepwise eye movements.

The desert ant *Cataglyphis bicolor* uses the sun, and particularly sky polarization, to navigate.

Limitations of fixing and tracking

Visually fixing and tracking an object are important orienting mechanisms that serve compasslike functions. Because those we have analyzed so far are merely short range and direct in action, their relevance to global navigation is rather limited. Yet they are important for the insight they yield on how visual data may be used to steer courses. Two other kinds of visual direction finding are obviously of great use to animals on a much larger scale. These correspond to techniques that are critical for human navigators, namely *piloting* and *celestial navigation*.

To pilot, an animal tracks an extended visual landmark such as the course of a river or a coastline to guide itself far beyond its original line of sight. Certain birds do this over long distances as described in Chapter 1. Or the navigator uses a recognizable sequence of discrete geographical details as successive fixes leading to the goal. Such direction finding from, for example, cape to peninsula to island requires recognition of specific landmarks and knowledge of their sequence. The topic of piloting by animals will be postponed until Chapter 8, which deals with the sense of space and the possible nature of animal maps. The rest of this chapter takes up visual direction finding dependent on celestial compasses of several types.

THE SUN COMPASS

As the sun, the moon, the stars and the sky are involved, a celestial compass is about as grand a navigational mechanism as earthbound animals could have. Because celestial reference points permit straight-line courses to be steered for long distances, such a direction finder offers guidance for longer trips than can be navigated by piloting or by fixation on or pursuit of nearby objects. Human navigators, whether primitive or sophisticated, employ celestial compasses. While a number of clues including stars are used for such direction finding by many animals, the sun is no doubt the primary one and was first discovered to fulfill this function in the desert ant *Cataglyphis bicolor*. This ant is a hunting species that needs no obvious terrestrial landmarks to relocate its home from ranges far beyond sight of the nesthole. Its ability to home rapidly and accurately from any direction across flat, featureless terrain relies on the sun for a reference bearing. This was proved first by using a screen to block the insects' vision of the sun, which disrupted their straight homeward path. Then when a mirror was so placed that it changed the apparent bearing of the sun, the ants shifted their course by exactly the same angle as the sun's bearing was changed by the mirror's reflection. In other words the insects were fooled into steering in the wrong direction by an artificial relocation of the sun.

Experiments with another ant, *Lasius niger*, added evidence that such insects steer a straight course from the sun. An ant's trip home was interrupted, and the ant was kept under a small box for 2½ hours. When uncovered it resumed a straight course again but in a geographical direction displaced clockwise by 37°, the same angle by which the sun's bearing had changed while the ant was under cover! This confirmed use of a solar compass but also raised a question that was not to be answered for several decades. Namely, as the sun moves through the sky at a rate of about 15° per hour, how can such a compass work for more than a matter of minutes? However, ignoring this question for the moment, we can fairly say that errors due to apparent solar motion are trivial for short time intervals.

Honey bee behavior

Many scientists have contributed to our understanding of direction finding in animals, but Karl von Frisch stands out both for his own spectacular discoveries and for his stimulating influence on students and colleagues. Von Frisch, Professor at Munich University in Germany, used honey bees *Apis mellifera* as subjects in many of his experiments. He would train worker bees to feed at any particular place up to several kilometers from the hive by putting out a saucer of scented sugar water for them at that spot. One of his students used this method to carry out an experiment on flat terrain with no obvious landmarks. He showed that if bees were detained at the feeding place, say for an hour, their course back to the hive was in error by a substantial clockwise angle. This displaced orientation, like that of *Lasius,* closely matched the angle by which the sun's bearing had changed while the insects were detained. Actually both these early ant and bee experiments are somewhat misleading in the light of later work to be discussed, but they introduced the basic idea that animals can find their way with a *solar compass.*

Among many other things, von Frisch showed that honey bees apparently learn by experience to use such a celestial direction finder. Worker bees employ different navigational techniques when they are inexperienced than they do after learning to home. *Naive* workers apparently use landmarks and other terrestrial clues to return to the hive. If they are displaced, for instance, to the west of the hive from a feeding place 150 meters south of it, they nevertheless home successfully. In contrast, *experienced* workers upon being displaced ignore landmarks and rely only on the solar compass. Consequently they miss the hive by a distance equal to their initial displacement. This and much subsequent work show that bees, as well as many other kinds of animals, use the sun to find directions.

However, depending on a compass in which the reference bearing may rotate clockwise through 360° over a single day as in the arctic summer or suddenly, as at the equator, switch at noon from steady east all morning to steady west all afternoon is rather daunting. Actually the sun's apparent path through the sky is a complex function of the latitude of the observer and the season of the year. The solar course can be described by two variables. One is its altitude, which is the vertical component of the path relative to the horizon. This is zero at sunrise and sunset and reaches a maximum at local noon. There seems to be little evidence that animals use this dimension of the sun's path for direction finding. The second variable is the

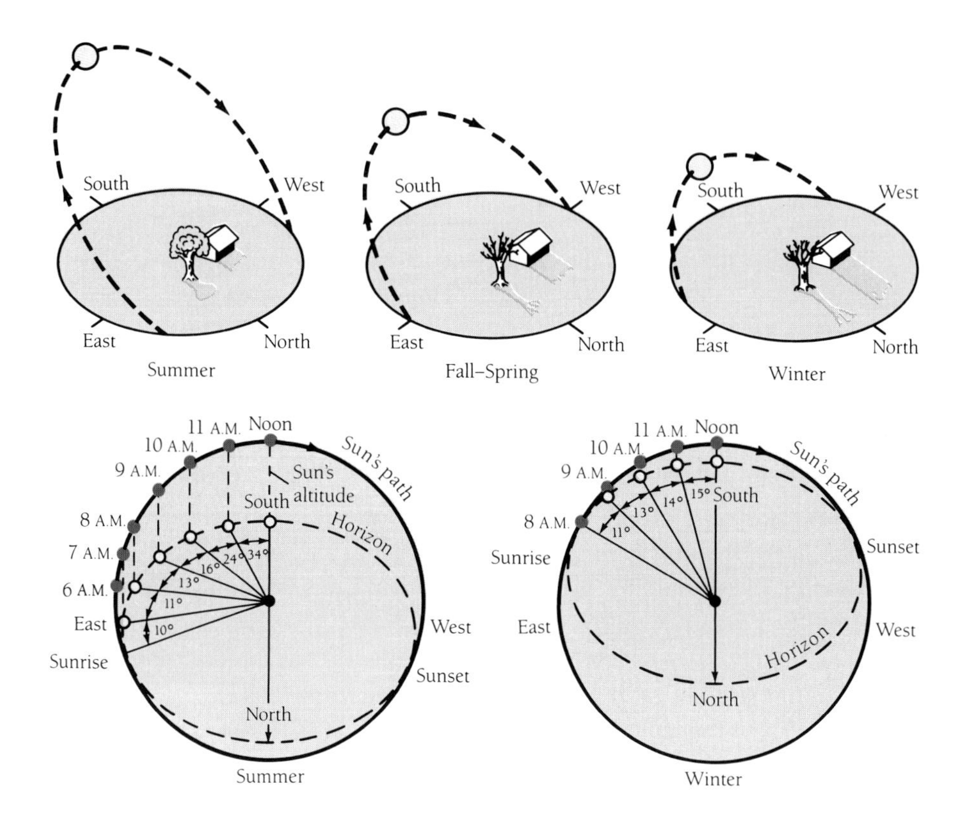

(*Top*) The effects of season on day length (elliptical path that traces the sun's movement), the location of the sun's rising and setting, and the highest elevation angle to which the sun ascends in the south at local noon for a location in the north temperate zone. (*Bottom*) Changes in the sun's bearing (azimuth) during the day differ depending on the season and latitude. Animal sun compasses respond primarily to the sun's bearing and little, if at all, to its elevation.

horizontal component of the movement, which is essentially the downward projection of the sun's location in the sky onto the observer's horizon. At any moment this is the *azimuth,* or bearing, of the sun, usually measured as a clockwise angle from north, which is taken to be 0°. Most evidence indicates that despite large changes during the day, this azimuth is the major—or only—reference for animal sun compasses. It was discovered, just before 1950, that animals can steer a steady geographic course from the sun over long periods. To do so they must allow for its continuous horizontal movement through the sky. This capacity was first proved independently for two quite different animals—the honey bee and the starling. Subsequent research has shown that a large number of animals possess such a *time-compensated* solar compass. Usually the sun's azimuth does not move at a uniform rate around the horizon. Rather, its rate of change differs with time of day, latitude, and season. How do animals make the necessary compass corrections?

Time compensation: bees

Von Frisch first proved that honey bees do correct their solar compass for the sun's daily changes in azimuth with the following experiment. A bee colony was trained to feed at a scented feeding station 200 meters to the west of the hive and then the whole hive was carried during the night to a distant spot with an entirely different terrain. The next morning four feeding stations were set up 200 meters north, east, south, and west of the hive. Of twenty-seven foraging bees arriving at one or another of the stations, only seven bees altogether went to the north, east, and south feeding places, but twenty came to the one in the west. Thus most bees flew out from the hive on the geographic course previously learned. Because of the unfamiliar hive location, the solar compass appeared to be the only direction finder available to them. The last flights to the west on the previous afternoon had been steered from the sun's azimuth in the southwest. On the next morning it was in the southeast. Corrections for substantial differences in the sun's azimuth with time of day must be routine for honey bees.

In addition to such feeding experiments, von Frisch and his students also used social behavior of honey bees within the hive to learn about the insects' celestial navigation. Among other things, scout bees inform their hive mates about the location of food by repetitive sequences of movement that von Frisch called dances. These ritualistic patterns, performed on the honeycomb surface, communicate both the *bearing* relative to the sun and the *distance* of the flower patches concerned. Except when these food sources are within 100 meters of the hive, bees perform a so-called waggle dance. This traces out a figure-eight motif and is repeated often in the same orientation. The insect begins each phase in the dance by walking a short straight course, followed by half circles alternately to the right and left back to start again. The initial straight walk, during which the bee vigorously wags its abdomen and emits bursts of sound, points directly toward the food source, provided that the dances occur on a horizontal comb surface with a clear view of the sun. The correct orientation and accurate repetition of this figure eight are lost if the bee's view of the sun's azimuth is cut off by a screen. Actually under normal conditions inside the hive, bee dances take place in the dark with no view of the sun or sky and on a vertical comb surface. In that case how can the bee dances still show the direction to a flower patch? This feat depends on taking *up* (sensed by gravity perception) as the solar bearing and orienting the straight element of the dance to the right or left of up by an angle equal to that seen by the bee in flight between the sun's azimuth and the course from the hive to the flower patch being signaled. Such a transfer between a solar reference and a gravity reference is something like our convention of orienting geographic maps so that north is up. An additional feature of the bee dance, namely, its tempo, signals the distance of the goal from 100 meters out to 10,000 meters or more. This slows down with increasing distance of the flower patch so that the rate of performing the figure eights decreases while the number of waggles in the initial

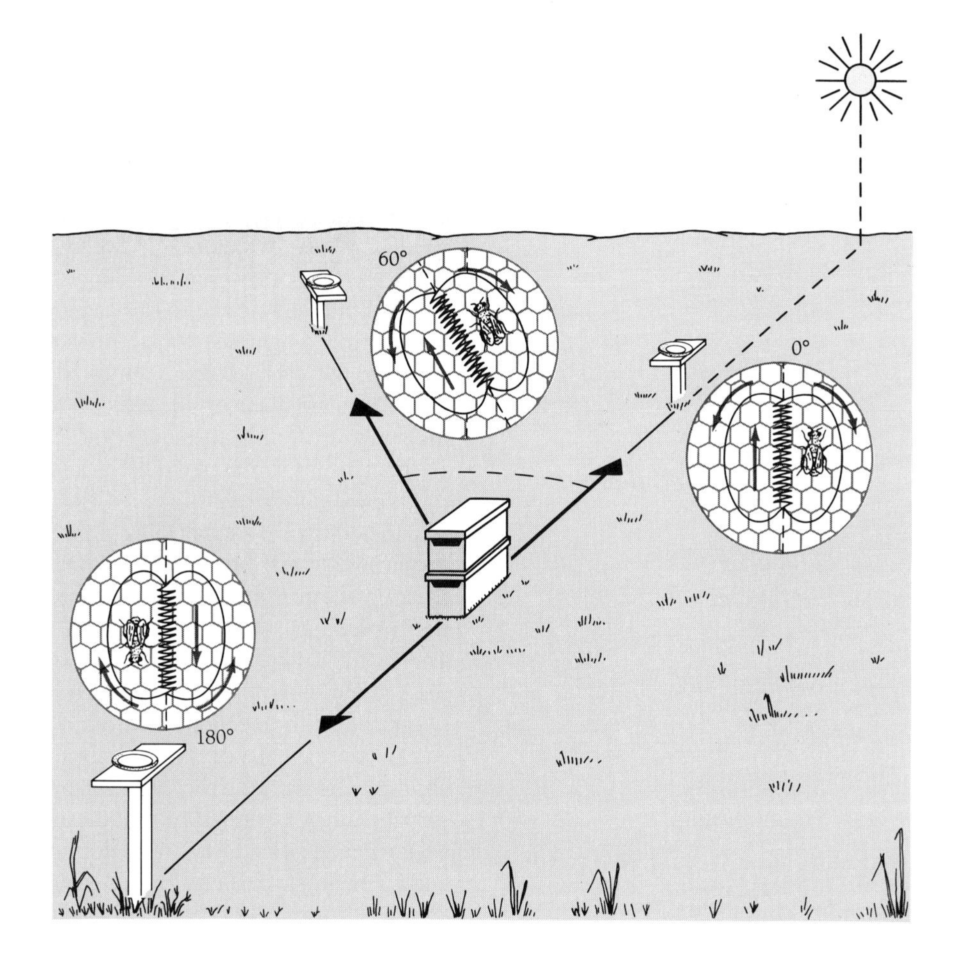

The waggle dances of the scout bees communicate the direction and distance of food sources to their hive mates. When the direction of the food source is the same as that of the sun, the waggle run points straight upward in the vertical honeycomb (0°); when the site is 60° to the left of the sun, the waggle run is oriented 60° to the left of up; when the food source is 180° from the sun's bearing, the run is pointed straight down.

straight segment increases. Not only can fellow bees read and interpret these movement patterns, but human observers have learned to do so, too. Consequently bee dances provide detailed insight into this insect navigator's methods.

Usually brief, the dances of scout honey bees may sometimes signal one goal for some hours. Then their ability to correct their solar compass for the sun's changes in azimuth can be directly followed. The bee's dance orientation is constantly adjusted so that it points correctly toward the target despite the sun's considerable change in bearing during the period of observation. To do this the bees must know the direction and rate of the sun's movement along the horizon and must measure time. How, then, would a northern-hemisphere bee use its solar compass if it were flying in the southern hemisphere? The problem here is that north of the tropics the sun always

A pollen collector (center) performing a waggle dance.

appears to move *clockwise* from east to west. It reaches its highest elevation when it is exactly *southward*. South of the tropics the sun appears to move *counterclockwise* from east to west and is at its zenith and directly *north* at noon. Within the tropics the apparent rotation changes from season to season. I remember myself feeling a bit disoriented in Canberra, Australia, which is in the south temperate zone, when I first noticed that the sun seemed to be moving counterclockwise rather than clockwise. How about the bees?

Experiments show in fact that worker bees become disoriented when transported far enough north or south to reverse the apparent solar movement. It may take them several weeks to learn the locally correct path of the sun. For instance a bee colony trained to feed north of the hive in Kandy, Sri Lanka, where the sun appeared to move counterclockwise was transported overnight to Poona, India, where the solar path was clockwise. There the displaced bees flew south from their hive to look for food. This was good evidence that only the sun compass was being used for this course finding even though north and south were reversed. Displacement experiments between North and South America generally confirmed these Asiatic results. Other animals, including the beach sandhopper, fish, homing pigeons, and Adelie penguins all behaved similarly when experimenters transported them from their native hemisphere to the other one. Clearly, arctic terns and other birds that annually spend extended periods at high latitudes in both hemispheres must regularly cope with this problem.

Time compensation: birds

If caged wild birds of migrating species can see the clear sky, they tend during the fall and spring migration periods to move and flutter in the very direction they would fly if they were free to do so. Discovered by the German ornithologist Gustav Kramer, this migratory restlessness, as it has been called, has proved to be an enormous help in analyzing bird navigation. For instance, that birds have a solar compass could be readily confirmed by observing caged birds able to see the sun's bearing directly or experimentally displaced by a mirror. They were able to keep on course within 3 to 5°. The duration and intensity of migratory restlessness in various species are directly correlated with the length of the actual trip normally ahead.

Caged birds that can observe the sun's bearing readily learn to feed in a particular geographical direction. This is much like the feeding-training technique for honey bees. Starlings were held in a circular cage with twelve feeding cups spaced around its edge. First the birds were trained by rewards to expect food only in the cup toward the south, for instance. They quickly reduced the effort spent in searching for food in the other eleven locations. Using their time-compensated solar compass, they maintained a steady southward orientation all day while their azimuth reference moved through an angle of 180° or more.

Artificial sun compass

Kramer also showed that captive starlings in a laboratory will accept a fixed light bulb as a stand-in for the sun. In such a case the birds, instead of maintaining a fixed course to a moving reference as in nature, follow a systematically changing course steered from a fixed reference. Hence simple stereotyped responses are important components of complex behavior like navigation even in higher vertebrates. Honey bee workers also will use a small laboratory light source as a directional reference for waggle dances. Their geographic dance orientation on a horizontal surface can be changed at will by illumination with a spot of light from various bearings. In such experiments major differences in the vertical height of the artificial light source have negligible effects on orientation behavior. In contrast even minor horizontal light displacements evoke corresponding figure-eight rotations. This supports much field evidence that the reference bearing is by far the major factor in the solar compass.

Sun compass in fish

Celestial navigation is not limited to aerial and terrestrial species. Fish, like birds, mammals, amphibians, reptiles, crustaceans, spiders, and insects, also use a solar compass. Sunfish *Lepomis gibbosus* and *Lepomis macrochirus* learn to feed in a southerly direction or to escape to the north when held in a shallow tank open to the clear sunny sky. Like that of honey bees and starlings, the sunfish solar compass is time compensated, and an artificial light source can serve as a surrogate sun.

Solar-compass malfunction may explain the failure of attempts to transplant sea-going, self-sustaining salmon populations from the northern to the southern hemisphere. Repeated, large-scale attempts to introduce Atlantic salmon to New Zealand, Tasmania, and the Falkland Islands have failed. Except for landlocked populations, the only successful salmonid transplants have been some sea trout introduced to the Falkland Islands and a somewhat limited population of chinook salmon (locally called quinnat) to certain rivers of the east coast of South Island, New Zealand. Because the sun's apparent movement through the sky is reversed in the two hemispheres outside the tropics, a northern hemisphere fish relying strictly on a solar compass appropriate to its origins could well be unable to navigate effectively. Some evidence that solar-compass courses are genetically fixed in fish would support this hypothesis. However, two other explanations have been suggested: According to one of these, cool, extensive river systems comparable to the Fraser and the Columbia are absent in the temperate southern hemisphere because of its smaller land masses. According to the other explanation, circulation patterns in the oceans of the southern hemisphere, differing as they do from those of the animals' original habitat, may hinder the return of adults to the mouths of their natal streams.

Obviously these attempts to account for limits in the adaptability of salmonids are rather speculative and have led us somewhat far afield from the sun compass, our major point of interest here.

Time and the sun compass

To correct the solar compass for the sun's movement, animals must depend on their internal sense of time. This connection between compass and clock was directly demonstrated in caged starlings trained to feed in a certain direction under the natural clear sky. After being trained the birds were taken into the laboratory and subjected to an artificial day having both dawn and dusk delayed by 6 hours. Over a period of several days this new light-dark schedule shifted the birds' time sense to match the laboratory day. Recall that during the 6-hour period of this time change the sun appears to move through the sky by about 90°. Will this difference between solar time and clock-shifted bird time have some effect on navigation? Put out to feed under the natural sun and sky, the clock-shifted starlings, indeed, chose to feed 90° clockwise from the trained orientation. Two other groups of starlings were tested to confirm this relation. One group, kept in the laboratory with artificial days that advanced its time sense by 6 hours, searched for food under the natural sky 90° counterclockwise from the trained heading. Another group of birds, the "controls," that saw only the natural sky during the whole experimental period continued to feed throughout in the direction originally learned. These results prove that resetting

A star chart showing the pattern of stars in the night sky as seen from one celestial pole to the other at about 90° longitude. Even when using the same celestial clues as humans, animal navigators appear to depend on the bearing of the sky elements rather than on their angle of elevation.

the internal clock of the animal is equivalent to displacing its solar compass by an angle that matches the sun's movement during that time interval. Later, time sense was shown to affect homing pigeons in the same way that it did starlings provided that the sun is shining when the pigeons are finally allowed to see the sky.

Yet surprisingly, under heavily overcast skies, pigeons whose time sense had been shifted homed in the correct bearing to their loft! This proves that without the sun or clear sky the birds can still navigate using some other course finder. Whatever this is, it must be independent of the birds' internal time sense. In practice, some limitations of the sun compass require that there be alternative methods of finding the way. The sun even in clear weather obviously is not available for reference at night, which lasts for months during polar winters. Depending on latitude and season night prevails from 0 to 100 percent of a day's 24 hours. In some parts of the world and in some seasons, skies are heavily overcast for long periods during which no celestial clues can be seen for course finding. During clear weather the azimuth of sunset has been shown to be important for nocturnal migrators, which include most temperate and high-latitude song birds. The evening twilight period is a doubly important one for these navigators. During this time the birds seem to decide whether or not they will migrate during the coming night. Passerines in general are strongly influenced by current weather conditions in initiating their fall movements. On some nights few, if any, take flight, whereas on others massive numbers depart. Indeed we saw that some "dare-devil" overseas migrations in the western North Atlantic could only succeed if they are closely linked with just the right meteorological conditions. Hence twilight is significant for making the go, no-go decision as well as for preselecting the migratory course from the sky or from the bearing of sunset before darkness descends.

THE STAR COMPASS

During the night and at twilight, stars could serve as celestial direction finders. The so-called fixed stars, which include all those beyond the solar system, have a rather simple nocturnal course through the sky, particularly as observed near the equator, and thus provide important guidance for navigation. We have already seen that canoe navigators in the central Pacific depend strongly on a star compass for their interisland navigation. In the tropics a star that rises straight up from the east passes directly overhead through the zenith and sets in the west exactly opposite its place of rising. Such rising and setting points of recognizable stars provide excellent compass orientation. In contrast, stars seen from places near the earth's poles do not appear to rise or set at all but instead revolve in semicircular paths just above the horizon. Because all navigationally significant apparent motions of the remote (and therefore seemingly fixed) stars are due exclusively to the earth's rotation, the centers of star

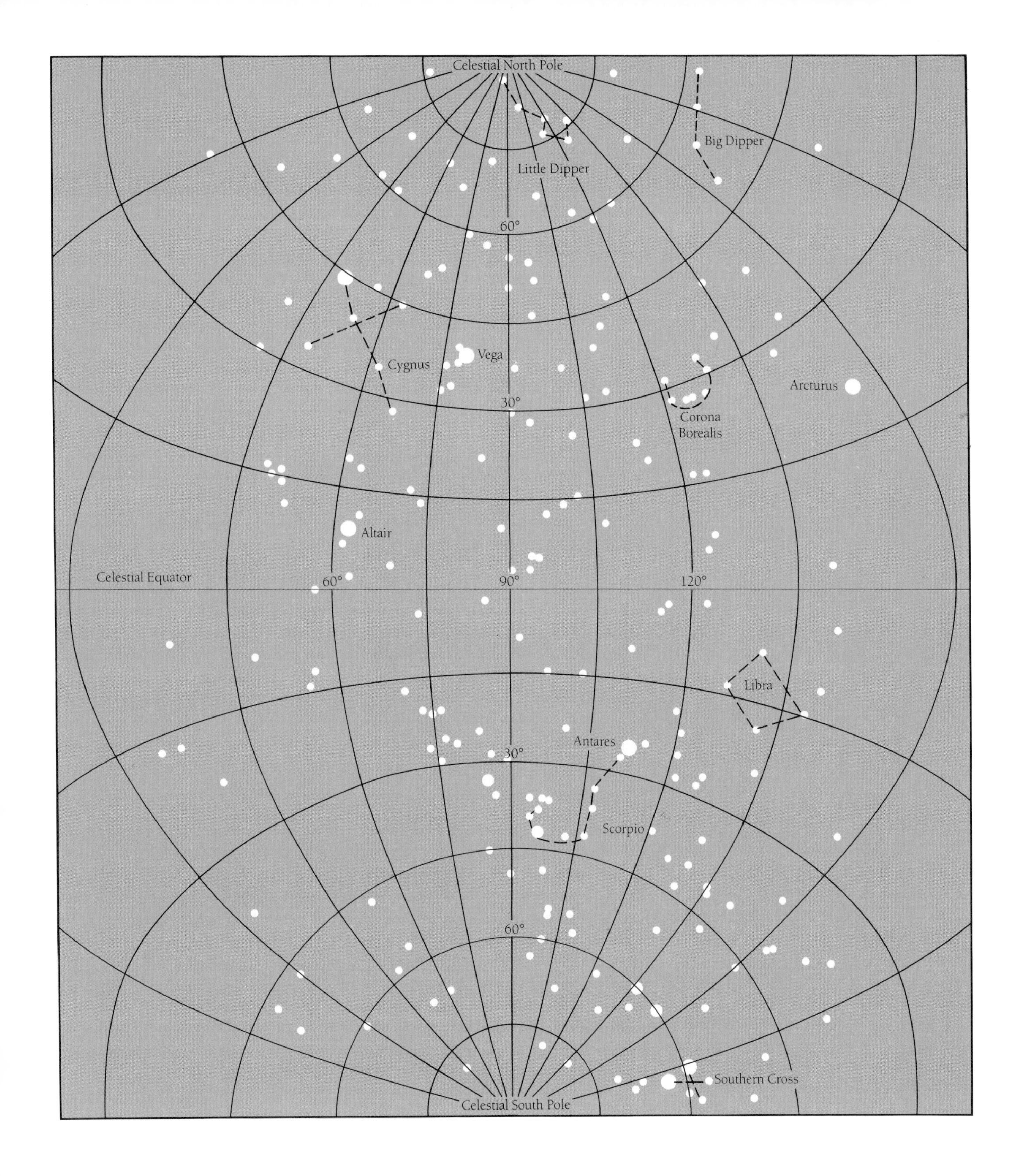
Celestial North Pole
Big Dipper
Little Dipper
60°
Cygnus
Vega
Arcturus
30°
Corona
Borealis
Altair
Celestial Equator
60°
90°
120°
Libra
Antares
30°
Scorpio
60°
Southern Cross
Celestial South Pole

paths in the Arctic and Antarctic are the celestial poles, points in the sky directly over the north and south geographic poles. In the northern hemisphere the celestial pole is rather closely marked by Polaris, the polestar. In the southern hemisphere it is identifiable as the center about which the prominent Southern Cross constellation appears to sweep each night like the hour hand of a 24-hour clock. These markers provide primary compass references for north and south respectively; other nearby individual stars are of little importance except to confirm the polar point from their rotation around it.

As seen in the temperate zones, the paths of some stars circle but the paths of many others rise in the east and move obliquely across the sky to set in the west. For sophisticated human navigation, including that in outer space, fewer than 200 stars are suitable for navigation of which fewer than 60 are ordinarily used. These are identifiable from star charts and to experienced observers from their locations. Because the earth revolves around the sun, the particular part of the celestial sphere visible at night varies with the season; the constellations of the zodiac, for example, appear to progress across the sky in succession. Consequently navigators using the stars must consider place, time, and season of observation. But before venturing further into navigational astronomy, we need to ask how is it relevant to animals. Migrating birds do use stars to find and maintain courses. For instance, migratory restlessness of caged birds with a view of the clear night sky is reasonably well aligned with the normal flight direction. On overcast nights such behavior is randomly oriented. However, migrating birds may maintain straight and accurate courses in the field under complete overcast. Then other orientation clues such as wind direction, sound patterns, the earth's magnetic field, and inertial navigation must take over.

Planetarium orientation

A planetarium was used to prove that birds can find direction for migration from the stars. Warblers about to migrate were brought into a planetarium whose star pattern was like that outdoors. The birds oriented just as they would have under the real night sky. If the pattern of stars was displaced, the birds shifted their orientation by the same angle. The planetarium sky seemed to provide the course needed for nocturnal migratory navigation. Note that the course chosen in the spring is opposite, or quite different from, that chosen in the fall even though the stars are used to steer both. Recall that an animal's migratory goal must be set independently of the navigation system. Hence seasonal differences in the preferred orientation of migrating birds must arise from *within* the animals. This has been clearly proved by experimentally shifting the birds' spring and fall physiological states. In nature increases in day length during the spring and its decreases during the fall stimulate the physiological changes. Monitored by photoreceptors coupled with the animal's hormonal system, day-length changes control preparation for seasonal activities such as migra-

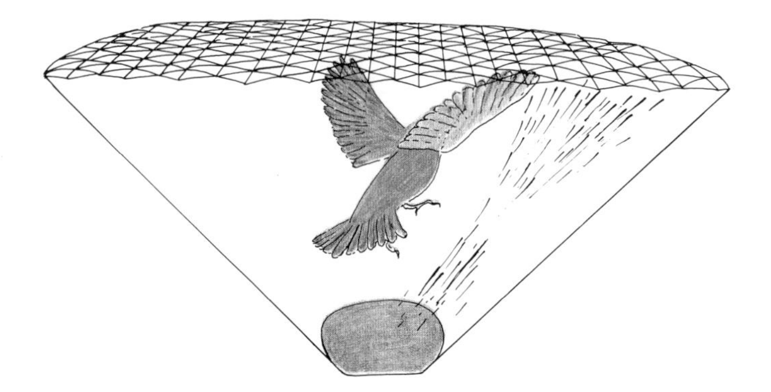

(*Top*) Experimental evidence for star navigation by migratory birds was obtained using young indigo buntings exposed to a planetarium sky. (*Bottom*) One method of recording the preferred orientation during a bird's migratory restlessness in planetarium experiments was to place the bird in a funnel that had an ink pad at the bottom and sides lined with blotting paper. (A series of such funnels is seen in the photo above.) The bird stepped on the ink pad and its flutterings in response to navigating by the planetarium sky left a pattern on the blotting paper that showed its orientation.

tion, molting, and reproduction. In the laboratory migratory restlessness, for instance, can be induced at any time of year, either by exposing birds to shortened or lengthened day lengths or by injecting hormones to alter the timing of their normal seasonal cycles.

Although the first planetarium experiments were received with some skepticism, later research, particularly on the indigo bunting *Passerina cyanea,* clearly confirmed that on clear nights migratory birds use a star compass to find their way. Birds reared in captivity are needed for such work because any previous exposure to the stars under natural or planetarium skies influences later behavior. Careful experiments showed that learned star patterns are used by buntings to choose geographic directions.

As time compensation is not involved, the star compass must operate quite differently from the sun compass of insects and birds. For the indigo bunting a sky area within about 35° of the polestar is most important, suggesting that the main compass reference is at or near the sky's north pole. Individual stars or constellations are not essential. Thus information in some parts of the star pattern is redundant. Yet blocking out substantial areas of the sky does decrease the compass's effectiveness. To use this stellar course finder the birds must first experience rotation of stars around the celestial north pole either in nature or in a planetarium. After a period of such learning, the birds can orient effectively to planetarium stars even when they are not moving. Indigo buntings without previous exposure to the natural sky can be disoriented in a predictable way by rotating the planetarium sky about some other star than Polaris. If that simulated sky rotation is transported to the west from its normal pole in the north, for instance, the birds' star compass is displaced 90° counterclockwise. Hence the apparent center of the circling stars does provide the bird's geographic reference.

Star compass in other animals?

Without doubt, some migratory birds do orient from the stars but there is little evidence for this ability in other animals. It seems rather unlikely that any aquatic animals do. Because water strongly absorbs and scatters light and its rippled surface also deflects light, stars are probably seldom visible to animals in the water. Whether the compound eyes of insects can see stars at all is dubious owing to the low ability of these eyes to perceive fine visual details. Nevertheless some experiments on nocturnal moths orienting under clear, moonless, night skies suggest that these insects may use stars to steer a straight course, but more quantitative data are needed. Also, recent experiments on the rock crab *Leptograpsus variegatus* imply that perhaps a dozen of the brightest stars should be detectable by the compound eyes of this crustacean under ideal conditions.

THE LUNAR COMPASS

The moon might seem a more likely celestial object for helping animals find their way than the stars, but there have been relatively few studies of how animals orient at night. The moon's apparent size and brightness make it easily visible for most eyes. Using the moon would not require discrimination of pattern or detection of rotation. On the other hand, the moon can only be seen for varying fractions of the

The beach flea *Talitrus saltator* uses celestial clues to move with the tides up and down the beach in a direction perpendicular to the shoreline.

night, which limits its use for navigation. Animals respond to lunar rhythms, particularly to those related to tidal influences in the sea. For instance feeding and other local movement in most intertidal animals are strongly correlated with the tides. Also, the full or nearly full moon may sometimes disorient nocturnally migrating birds from their proper path. This has also been observed in caged migrators showing spontaneous directional preferences. Yet nocturnal field data are inadequate to document the extent and effect of such distraction.

Course finding from the moon has been found, however, in a number of arthropods such as the beach flea *Talitrus saltator*. These small shore-living crustaceans and a number of other kinds of beach or riverbank animals, both invertebrate and vertebrate, commonly need to know the geographical direction perpendicular to their native shoreline. This allows them to travel quickly from water to land or from land to water. Also it allows the beach flea to move up and down the beach with the tide in order to stay within the narrow band of damp sand that is their preferred habitat. If it is too dry, they move down the beach; if it is too moist, they move up. During the day the sun serves to orient this movement as does the moon at night.

THE POLARIZED-LIGHT COMPASS

Von Frisch discovered another sky compass in the late 1940s while studying the dance orientation of honey bees. He noticed that they could maintain their bearings without the sun as long as they could see some area of clear blue sky. Even 10° of blue sky permits them to maintain the correct orientation. When the dance area in the hive was completely covered except for a small opening exposing the bees to a restricted patch of blue sky, the bees still retained a quite normal sense of direction. If, however, the sky visible through the opening was covered with clouds, the bees lost their bearings. The basis of their orienting to blue sky without the sun turned out to be the strong polarization of scattered sunlight in most clear sky areas. This polarization had not been discovered by western science until early in the nineteenth century, but there is some evidence that Viking navigators made use of it a millennium earlier. This might explain the reports that they depended on a certain crystal as a "sun stone" to read the solar azimuth during long voyages in subarctic twilight. Obviously, animal navigators had evolved their own neat method of using polarized skylight ages before.

Von Frisch proved the existence of such an optical compass by placing a polarizing filter over the outer end of the small aperture through which the dancers could see a bit of sky. By rotating the filter to alter the direction in which incident light was polarized, he could change the orientation of the bees' dances at will! Thus it was discovered in 1948 that the compound eyes of these insects can perceive the polarization of natural light and that this perception gives the insects an additional celestial compass.

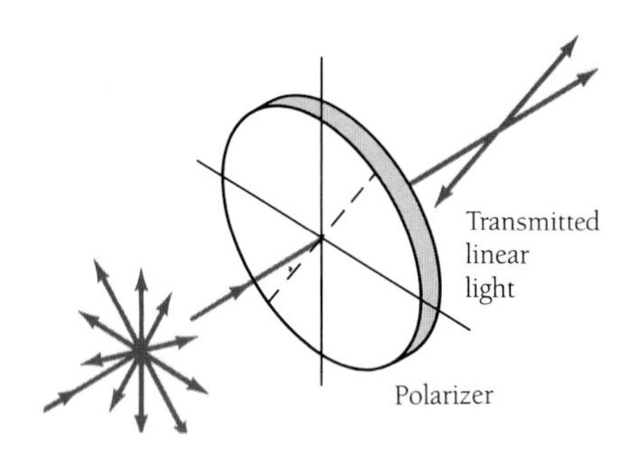

Unpolarized light enters a polarizer that blocks the passage of all waves not in its transmission plane. The reduced amount of light emerging from the polarizer is said to be 100 percent linearly polarized.

Sky polarization

Light consists of many electromagnetic waves that oscillate perpendicularly to their direction of travel. In direct rays of the sun the waves oscillate in all possible perpendicular planes around the transmission axis, and the light is therefore said to be not polarized. In fully polarized light the waves all oscillate in a single plane—the polarization plane, or *e*-vector. The light from the blue sky is partially polarized because more waves vibrate in one plane than in any other. Both the sky's blue color and the polarization arise in the atmosphere from the scattering of direct sunlight by air molecules and other fine suspended particles. Because the scattered sky light is polarized perpendicular to the original direction of travel of the solar rays, the whole blue sky has an *e*-vector pattern aligned like lines of latitude on a celestial sphere with the sun at one pole and the "antisun" at the other. Seen by an observer on the earth's surface, the amount of polarization varies from zero around the two poles of this sphere to a band of maximum polarization reaching 70 to 85 percent centered on its equator.

As the sun moves through its course from sunrise to sunset, the whole sky polarization pattern rotates with it. Two striking features that signal this movement are the "holes" in the polarization near the sun and the antisun and particularly the maximum polarization band at 90° to both of those points. As observed from near the equator, this band, which no doubt provides the basic compass information to animals that use it, runs north and south through the zenith at sunrise, rotates down toward the western horizon until noon, rises from the eastern horizon thereafter, and finally ascends to the zenith again at sunset. In temperate and polar latitudes the polarization movements are more complex and depend on the seasons as do the sun's paths. In brief the polarized-light compass extends the use of the sun as direction finder; rather than relying for reference on a solar image less than one degree in diameter, animals using this sky compass take directions from *e*-vector distributions spanning much of a celestial hemisphere. There are two problems that make the more extensive system less easy to understand than the solar version itself: How can the animal navigator perceive the polarization pattern? How is the sun's azimuth then deduced from that pattern? Neither of these questions have intuitively obvious answers; indeed they are the subjects of considerable recent research.

Perceiving the e-vector

One way to approach the question of perception is to analyze eye function in different species known to detect the *e*-vector of polarized light. Both the evidence for this detection and the visual mechanisms involved vary widely. For instance most humans can directly perceive *e*-vector orientations by close scrutiny of a slowly rotating polarizing filter against a distant uniform background. After some practice, even the

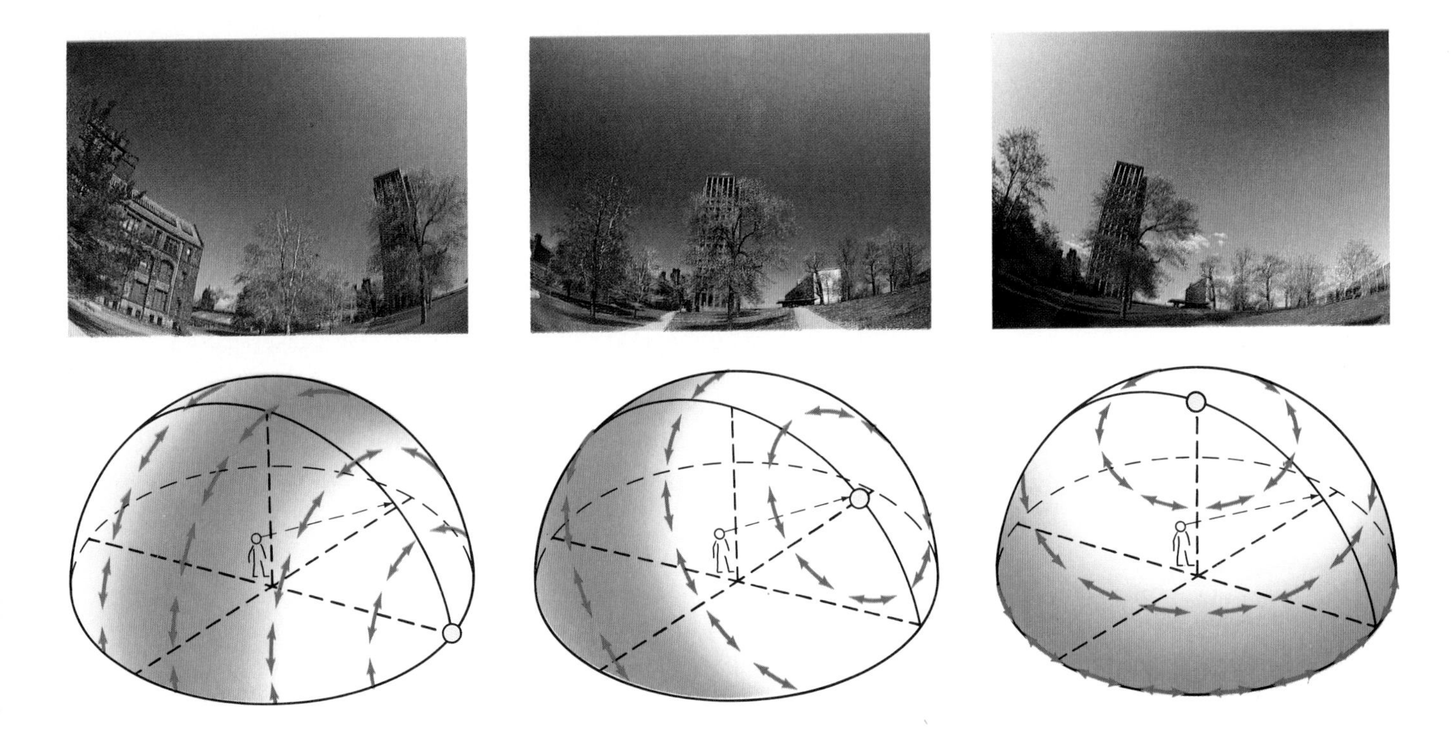

(*Top, left to right*) The band of maximum sky polarization seen as a dark strip at 9:00 A.M., noon, and 3:00 P.M. (*Bottom, left to right*) The pattern of sky polarization with the sun on the horizon, at 45° elevation, and at the zenith. The *e*-vectors are depicted by double-headed arrows and the degree of polarization by blue shading.

Many insects, like this leafcutter bee, have three tiny simple eyes, called ocelli, in addition to their compound eyes. (The ocelli are visible as the three dark dots just behind the antennae.) Although the function of these accessory visual organs is not well understood, in certain species they serve in polarized light orientation.

partial polarization of the sky can be sensed when it is very clear. The small, faint blue and yellow image perceived by most people who look for it carefully is called Haidinger's brushes and appears oriented according to the *e*-vector direction of the stimulus. This nearly subliminal phenomenon apparently depends on a yellowish pigment near the center of the human retina. By observing Haidinger's brushes in the sky, one can directly confirm major features of the *e*-vector pattern. Experiments have shown that particular species of fish, amphibians, reptiles, and birds orient to the plane of polarization in the sky. In the laboratory selective orientation to artificial polarization patterns has also been shown in a number of vertebrates. Direction finding from polarized light in nature is thus a potential navigational aid. But substantial evidence for its use by the long-range migrators in this major group is largely lacking except for some supportive behavioral experiments on juvenile sockeye salmon.

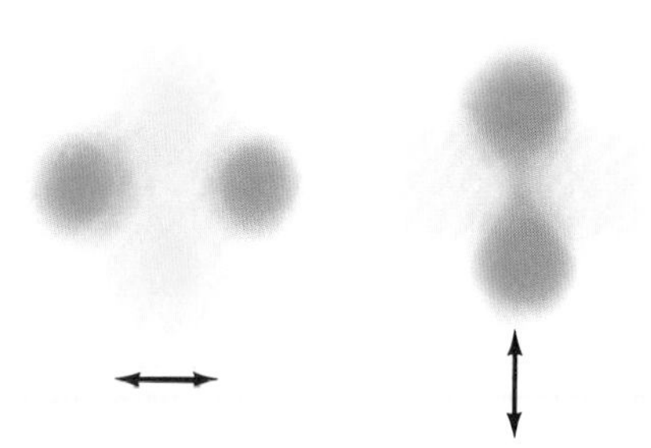

Haidinger's brushes. The blue sections are parallel to the *e*-vector, which is depicted by the double-headed arrows.

In addition, the sensory mechanism underlying vertebrate orientation to polarized light is not yet properly understood. The yellow pigment responsible for our perception of Haidinger's brushes occurs only in primates. To detect polarization more generally, some elements in the visual system must respond differently to

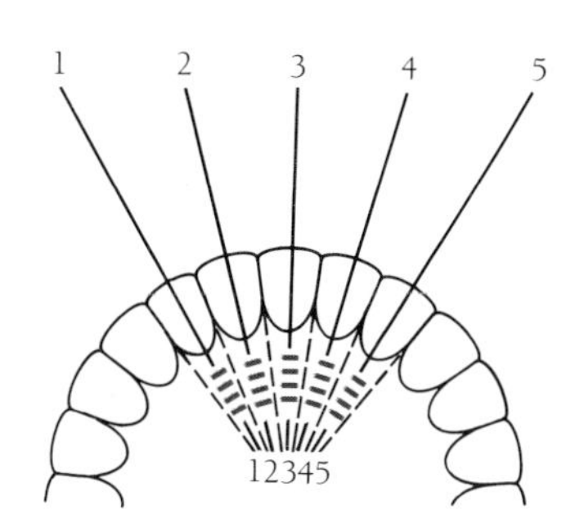

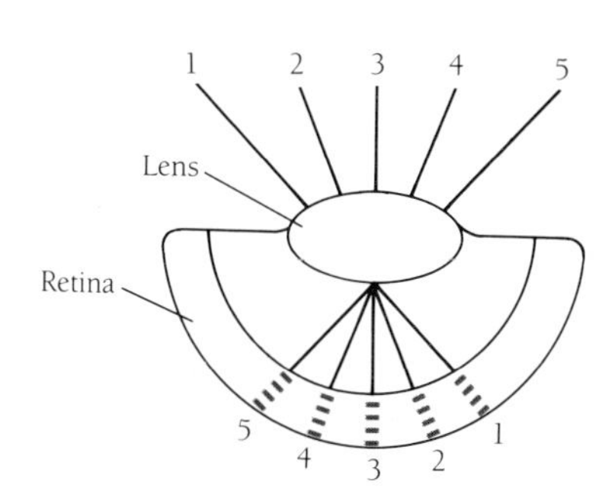

(*Left*) What a human with compound eyes and a housefly with camera eyes might look like. (*Middle*) A compound eye has many lenses and other optical elements that illuminate small clusters of photoreceptor cells. As depicted by the numbered lines, the overall image is convex, erect, and not reversed. (*Right*) A camera eye has only a single lens and optical system that focuses an inverted and reversed image onto the concave retina.

e-vectors oriented in various directions. In an extreme case one unit would not respond at all say to a horizontal *e*-vector but would react maximally to a vertical one; another unit might also respond selectively but maximally to a horizontal *e*-vector and not at all to a vertical one. In the retina of vertebrates, however, the rods and cones, which are the light receptor cells, are not known to respond selectively to the polarization plane of light naturally focused on them. Yet, in some of our research on goldfish, nerve cells in the visual area of the brain showed high sensitivity to *e*-vector direction despite uncertainty about how such discrimination could originate in the eye itself. Recent research has confirmed *e*-vector perception by goldfish but has not yet clarified how the feat is accomplished. In contrast to this still rather unimpressive understanding of both the mechanism and practical application of *e*-vector discrimination in vertebrates, knowledge of these functions is significantly more satisfying in some other animal groups, especially insects. A wide variety of eye types are included, ranging from minute ocelli of insects, mites, and spiders to large compound eyes of insects and crustaceans (all of which are arthropods) as well as the well-developed camera eyes of squid and octopus (which are mollusks). A good question here is: What common feature enables such apparently disparate visual organs to share the ability of strongly discriminating the *e*-vector?

It turns out that all these invertebrates that easily perceive the *e*-vector of polarized light share a basic feature in the fine structure of their photoreceptor cells. This allows them to respond selectively to *e*-vector direction. Light-sensitive pigment molecules are essential for vision and are always closely packed in special areas of the surface membranes of retinal sensory cells. In turn these membranes usually are organized in a systematic way for light absorption either spread out in flattened disks, as in the vertebrates, or rolled up in tiny tubules stacked together by the thousands. A membrane system made up of such packed and parallel receptor tubules is the hallmark of arthropod and mollusk eyes that can discriminate the polarization plane. The system's ability to absorb light selectively depends basically on the *e*-vector sensitivity of individual visual pigment molecules made manifest by

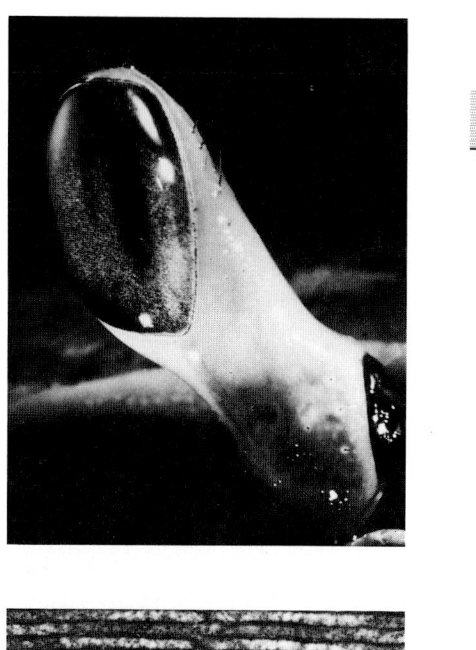

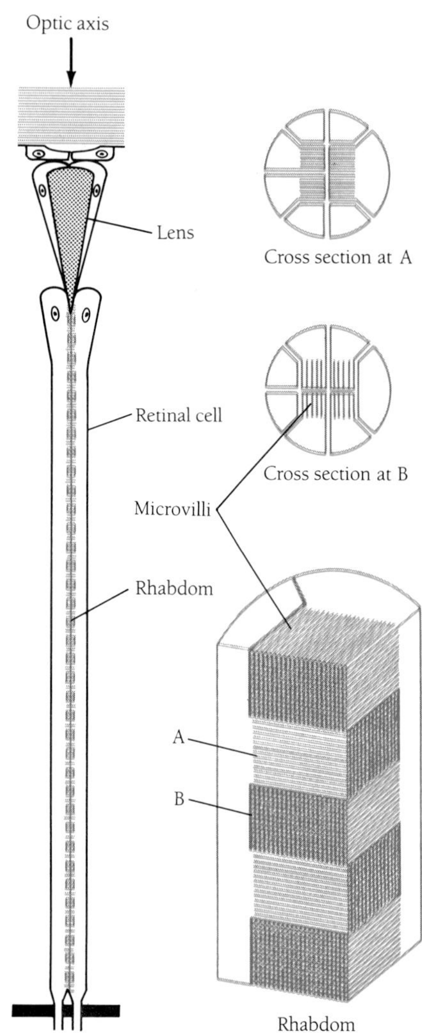

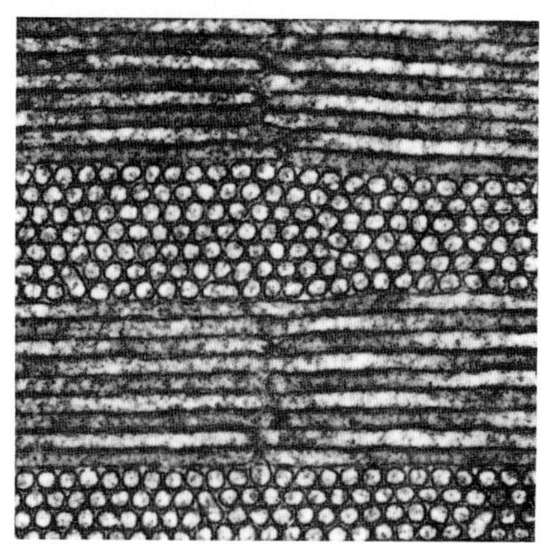

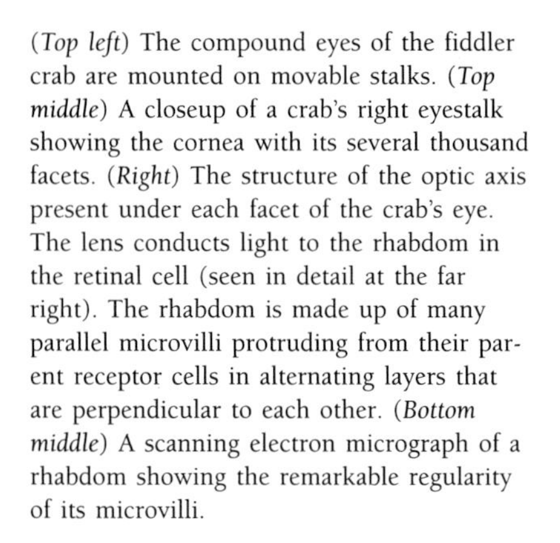

(*Top left*) The compound eyes of the fiddler crab are mounted on movable stalks. (*Top middle*) A closeup of a crab's right eyestalk showing the cornea with its several thousand facets. (*Right*) The structure of the optic axis present under each facet of the crab's eye. The lens conducts light to the rhabdom in the retinal cell (seen in detail at the far right). The rhabdom is made up of many parallel microvilli protruding from their parent receptor cells in alternating layers that are perpendicular to each other. (*Bottom middle*) A scanning electron micrograph of a rhabdom showing the remarkable regularity of its microvilli.

their regular aligned orientation within the receptor membrane and the close parallelism of all the tubules emerging from a given sensory cell. As a result of the almost crystalline regularity of this molecular and membranous array, such retinal cells are five to ten times more sensitive to light polarized parallel to the tubular subunits than they are to light with its *e*-vector at 90° to that axis. Usually in the eyes of crayfish, crabs, octopus, honey bees, and many related species, two sets of retinal cells lie with their major *e*-vector absorbing axes at 90° to one another. How this two-channel sensory input may be perceived by the animal is an intriguing, if moot,

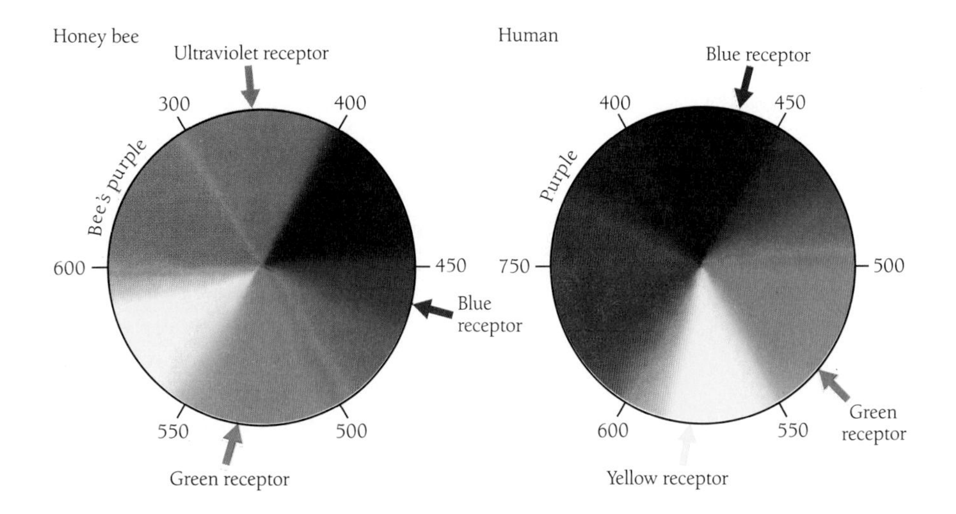

Honey bees and humans both have three types of color receptor cells, however, the bee's receptors are most sensitive at significantly shorter wavelengths than ours. Bees have an ultraviolet receptor with a peak wavelength sensitivity at 320 nanometers, which we lack completely; but they do not see toward the red end of the spectrum as well as we do. The bee's spectrum is shifted so that it contains ultraviolet, violet, blue, green, and yellow sectors in place of our violet, blue, green, yellow, orange, and red sectors. For both honey bees and humans a mixture of the two ends of their spectra produces a nonspectral color, "bee's purple" for the bee and purple for us.

question. Does the animal see color or black and white intensity patterns? Or does it perceive some special sensory signals for polarization never experienced by humans? A less abstract question more directly related to navigation is: How does the animal find its way from *e*-vector information?

The bee's sky map

To use the polarized-light compass an animal's perception of all or part of the celestial *e*-vector pattern must be converted into the geographic bearing of its goal. This conversion could be made by first determining the sun's azimuth from sky *e*-vector directions and then setting the chosen flight course from that just as with a solar compass. Estimating the sun's bearing from the perceived polarization is the novel step here. Does the insect have in its genes or in its central nervous system a complete detailed map of sky polarization at various times of day? Must the animal continually solve the complex calculations required to relate such map information to observed *e*-vectors in particular parts of the sky? Or can a straightforward built-in mechanism avoid much of this complication? Current understanding of this problem is best developed for the honey bee and substantially due to recent experiments by Rudiger Wehner, Sam Rossel, and their colleagues at the University of Zurich. Their data and analysis clearly support a rather simple, long-term solution to this navigation problem.

For their polarization compass honey bees, and probably a variety of other insects, need only a few receptor units looking up from the upper rim of the compound eye toward a rather small, elliptical area of zenith sky. About 140 out of a total of 5500 facets, as well as their associated internal optical and sensory components, are modified for a navigational function dependent on ultraviolet polarized light. Each of these special facets has a wider angle of view than others in the bee eye and is distinctive in having two opposite pairs of retinular cells with high polarization sensitivity. Their wavelength response is greatest in the near ultraviolet at about 350 nanometers, well below our violet-to-red visible range (400 to 800 nanometers). Because the sky is usually the only large-scale source of ultraviolet light in nature, Wehner has suggested that restriction to such short wavelengths permits ready discrimination between earth and sky. Interestingly the sun compass differs sharply in being insensitive to ultraviolet radiation but maximally responsive at blue and green wavelengths near the peak for direct sunlight.

The members of each pair of polarization-sensing ultraviolet receptors in the rim of the bee's eye complement each other in having their greatest responses to *e*-vectors differing by 90°. The differences in their outputs could strongly signal the orientation of a given *e*-vector perceived by that facet. The sequence of facets along the eye margin has the sensitivity axes of these critical receptors spread out in a fanlike distribution through 180°. Hence any given sky-polarization pattern evokes in both eyes a characteristic response pattern that depends on the angle between the animal's longitudinal body axis and the sun's azimuth. The specific *e*-vector direction in a given sky area determines which of the dorsal rim facets is most strongly stimulated as the insect scans the area with this eye region. Because of the symmetry of sky polarization on either side of a vertical plane passing through the sun and the observer, equal stimulation of the *e*-vector sensors in the margins of both eyes occurs when the bee heads either toward or away from the sun. If the full sky is clear and visible, discriminating the sun's bearing from the antisun's is no problem.

However, if the sun is hidden and only a fraction of clear blue sky can be seen, the problem becomes more difficult. Attempting to solve it, Wehner and Rossel have formulated a hypothesis as follows: To begin with, the double fanlike array of polarization sensors is likened to a template in which the eyes' fixed variation in the direction of the axes of *e*-vector sensitivity closely matches the mean distribution of the most strongly polarized *e*-vectors in the sky. As a result, maximum overall stimulation is perceived by this system when the bee's longitudinal axis is parallel to the sun's or antisun's bearing. Except at dawn and dusk the insect can select the sun's direction by choosing the alternative where the stronger polarization is perceived in the posterior half of the eyes' visual field. Some nontrivial compass errors are predictable for this simple mechanism particularly when the sun is high above the horizon and when only small patches of blue sky are visible to the bee. Although contested by some, the systematic occurrence of the errors expected from the hypothesis has been substantially documented by Wehner and his coworkers. Because we cannot see ultraviolet radiation and lack a patterned array of *e*-vector-sensitive receptors in our retinas, what the bee senses from the clear blue sky has usually been difficult to imagine. Yet accepting the Wehner and Rossel analysis provides a remarkable compass mechanism that may well account for this aspect of honey bee navigation. Their explanation implies that the polarization of light is not perceived as a distinct optical dimension but rather as a pattern of intensities or colors or both, useful for finding the way when its relation to the sun's azimuth is known.

Other insects

Desert ants, crickets, and other insects may have comparable built-in mechanisms for reading sky polarization and evoking the corresponding navigational behavior. Such a polarized-light compass needs a complex mechanism that apparently evolved only for a single important function—and for some species it has only a back-up function. On clear days bees preferentially use the sun for direction finding. When the sun is hidden behind clouds or by obstacles such as a tree or a mountain, they use the polarized-light compass. In contrast, the desert ant refers to this sky direction finder as its first choice for navigation. This remarkable insect also uses this *e*-vector compass as part of its accurate dead reckoning. Employing the polarized-light compass for determining direction along with some as yet unknown distance-measuring mechanism (number of steps taken? time spent running at a fixed speed?), this ant knows its position relative to its nest at any moment even on a long

The backswimmer *Notonecta glauca,* here suspended upside down from the water surface, has been shown in flight to use horizontal polarized light reflected from the surface of a pond as a landing signal.

wandering hunt for food. Although the outward path may be quite circuitous, the insect will immediately home accurately if alarmed. Clearly, it must perform some precise additions of course components to support such behavior. On such an emergency return trip, it uses sky polarization to steer back home.

Many other insects employ sky polarization in their navigation. The back swimmer *Notonecta glauca,* a water bug, shows quite specialized responses to ultraviolet polarized light reflected from the surface of ponds in which it lives. If such upwardly reflected light is horizontally polarized, it triggers a landing response when the insect flies over. The ultraviolet polarized light serves to identify a suitable place for the insect to alight. Dragonflies have ultraviolet receptor units in the *ventral* part of their relatively huge compound eyes that are effective for discriminating *e*-vectors. Although experiments are needed, such polarization analyzers might function as horizon detectors or position stabilizers for insects patrolling over calm water.

Underwater polarization

Polarization patterns underwater are primarily determined by scattering of directional light rays in the water. Such scattering establishes both the *e*-vector direction and the degree of polarization. Maximum polarization occurs at right angles to the light's direction of travel and may be as much as 60 percent in clear natural waters. In all lines of sight the resulting *e*-vectors appear perpendicular to the rays' direction. Hence, seen horizontally underwater, the polarization plane parallels the surface in the bearings of the sun and antisun, tipping maximally by 48.6° toward the sun's azimuth at 90° to these directions.

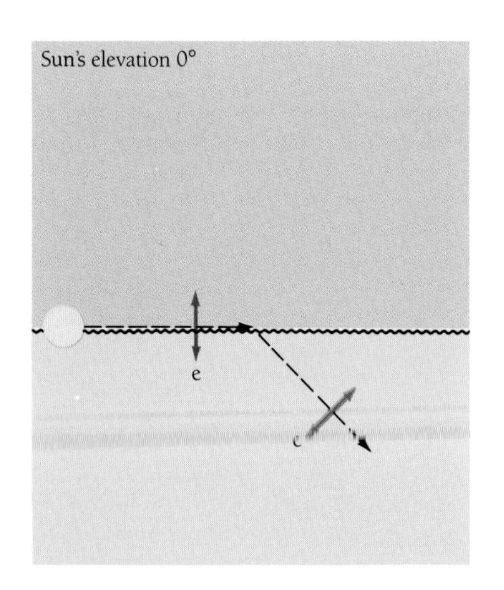

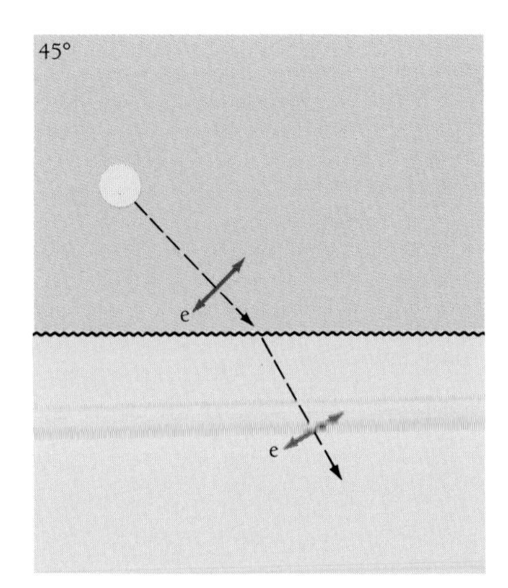

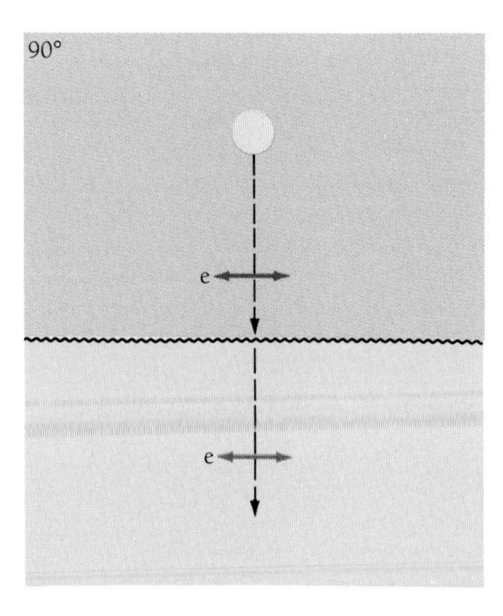

Underwater light polarization patterns depend on the sun's elevation and the observer's direction of view. (*Left*) When the sun's elevation is 0°, refraction at the water surface bends the rays downward at an angle nearly 49° to the vertical so that the *e*-vector appears, when viewed at 90° to the sun's bearing, to tilt toward the sun by that amount. (*Middle*) With the sun's elevation at 45°, the *e*-vector tilt is reduced to about 32°. (*Right*) With the sun at its zenith, there is no tilt.

Underwater polarization fills the whole visual field rather than only the upper hemisphere centered around the zenith for the sky. Also, as can be readily seen when snorkeling or scuba diving, the sun and the entire sky (including its polarization) can be observed from close under the surface. Because of refraction at the surface, opposite points on the horizon 180° apart in air appear compressed underwater to an overhead angle of 97.2°. The solid angle formed by the full horizon seen from the water is called Snell's window. Throughout the rest of the underwater field of view, polarization is present in patterns determined in the medium by the principles already outlined.

A polarized-light compass underwater?

Polarization underwater, like that of the sky, may provide a compass for navigating animals because the sun's bearing can be directly determined from the *e*-vector pattern. Unlike the sun's disk itself, the polarization pattern is visible far below the surface. The author, for instance, showed photographically that the sun's bearing was still clearly apparent from the *e*-vector orientations viewed horizontally at 200 meters depth near Barbados. This directionality should diminish at greater depths. Yet the polarization itself persists to the limits of visible light penetration (perhaps to about 1200 to 1500 meters in the clearest ocean water). While the solar compass would thus have been lost beyond about 400 to 500 meters, a stable pattern of horizontal polarization would be present to the limit of the photic zone into which sunlight penetrates. Even in deep water where the sun's bearing is no longer perceptible, polarization during the day would provide a fixed horizon reference for pelagic animals that can perceive it in otherwise near featureless surroundings. Most of them never see landmarks or even the water surface. Unfortunately experimental evidence to support such intriguing possibilities is not extensive.

The waterflea *Daphnia* was the first aquatic animal shown to orient specifically to the *e*-vector of polarized light.

We do know that many aquatic animals, including crustaceans, mollusks, and fish have effective levels of polarization sensitivity. In the laboratory they orient at fixed angles to *e*-vectors produced with polarizing filters. Yet the behavioral evidence so far obtained has yet to reach the sophistication needed to demonstrate the navigational usefulness of underwater polarization. This is partly due to the difficulty of carrying out the necessary experiments without grossly interfering with the natural underwater light pattern including the polarization. Also attempts to duplicate natural underwater illumination in the laboratory have been rather far from achieving the complex changes in *e*-vector orientation and degree of polarization present in various lines of sight in the sea. In addition suitable foraging, homing, or migratory behavior has not been found with which to study the polarized-light compass in animals as effectively as it has been in honey bees and desert ants. Further research is clearly needed to prove that the known *e*-vector-discriminating ability of many aquatic species helps their visual navigation. How important is such a compass in overcoming the handicap of severely reduced visibility underwater in the expert navigation of fish, porpoises, and whales?

6

Three More Senses for Direction Finding

The five traditional senses are sight, hearing, touch, smell, and taste. But this brief list, based mainly on rather naive human experience, fails to allow for the diversity of sensory information available to animal navigators. For example, the sensory information from statocysts, semicircular canals, and halteres (which, as we saw in Chapter 4, is important for navigation) is not apparent among the five traditional classes. These organs depend for their function on receptor cells that are sensitive to mechanical forces as are those for hearing and touch. A classification based on stimuli to which sense organs are specifically responsive is more effective. Six categories serve for such a classification: *Mechanical* receptors mediate touch, pressure, position, tension, acceleration, vibration, and sound. *Chemical* receptors respond to particular molecules and ions including those involved in smell and taste. *Light* receptors, which include eyes and other photosensitive structures such as skin, the pineal body, and parts of the midbrain, are sensitive to radiation at wavelengths between the near ultraviolet and the red. *Thermal* receptors respond to temperature in the skin or central nervous system; many animals have two types one for sensing warmth, the other for sensing coolness. Snakes and some other species have special organs for detecting infrared radiation from its heating effect. *Electrical* receptors, present in many aquatic vertebrates, are sensitive to natural voltage gradients in the environment and to electrical effects of muscle contraction in nearby animals as well as to electric signals generated by special organs in certain fish. Electrical receptors can also detect electrical changes caused by the movements of the fish themselves in the earth's magnetic field! *Magnetic* receptors, whose existence is taken for granted by many biologists, respond to weak magnetic fields such as that of the earth. To date magnetic receptor structures have been specifically located only

Salmon swimming upstream to spawn in their natal stream.

for certain bacteria and possibly certain birds. New sensory categories may, of course, be added to these six at any time. After all, electroreceptors were only discovered in the 1950s.

Being enclosed by a specialized receptor membrane is one of the basic features shared by all sensory cells. This membrane is exquisitely keyed to its particular special stimulus: light for example, in eyes; sound waves for ears; sugar molecules in chemoreceptors for sweetness; and so on. Only a few photons of light, a minute change in sound level, or a small number of sugar molecules are all that are needed for responses. The way stimuli act on receptor cells can be described in electrical terms such as *potential* (a difference in voltage), *current* flow (measured in amperes), and *resistance* (measured in ohms). The action takes place at the receptor membrane, which resists the flow of charged particles through it, but from time to time its resistance changes. Because of differences in the distribution of charged particles from one side of the membrane to the other, there is typically a voltage difference across it of 50 to 100 millivolts, the cell's so-called resting potential. In a computer circuit electrons are the charged particles flowing, but in living systems *ions,* such as sodium and potassium, behave in a similar way. In vision, for instance, light triggers a photoreceptor cell's response by reducing its membrane's resistance to ion flow. Sodium ions move through the membrane into the cell. This current evokes further changes that eventually transmit the sensory input from cell to cell toward the central nervous system.

A sensory cell, which may be either excited or inhibited by a stimulus, is not only specialized for a particular type of signal such as sound or light, but is also responsive to various stimulus dimensions such as intensity, frequency, or wavelength. The actual information transmitted to an animal's brain is an elaborate, coded barrage of nerve impulses. Nevertheless, relatively simple oriented responses can be evoked by sound or by chemical gradients in the environment or by other directional stimuli, making possible the different types of direction finding discussed here. Animals have evolved direction-finding mechanisms utilizing each of the six classes of stimuli, and compasses are known for perhaps four of these. In this chapter we will learn about animals' mechanoreceptive, chemoreceptive, and thermoreceptive direction finders and in the next chapter about their electric and magnetic counterparts (Direction finding with light was discussed in Chapter 5.)

Using their whiskers, rodents such as the *Peromyscus maniculatus* follow landmarks to adroitly navigate through darkened burrows. The series of contacts made by the elongated hairs along the burrow walls provide accurate steering control.

MECHANORECEPTIVE DIRECTION FINDING

A blind person feeling the way with a cane illustrates direction finding by touch as does a rodent tracking a burrow with its whiskers. Following landmarks by touch provides steering control that can be considered to be a sort of piloting. Only one or a series of oriented contacts is needed to go handily from one place to another.

Clearly an environment sufficiently rich in touch points, such as a branching tunnel or cave system, can be readily negotiated by adept species such as cockroaches and mice that are virtuoso maze runners.

Some animals can obtain information useful for navigation from movements of surrounding air or water. The antennae of flies and the bristles on the head of desert locusts, for example, can indicate wind direction and speed to the flying animals. The directional pressure of the wind bends and thereby stimulates these appropriately located mechanoreceptors. The animal must, of course, distinguish air pressure induced by its own flight from wind, which requires further input from vision about its movement over the ground or from a highly sensitive inertial system about its displacement. In contrast to fliers and swimmers *walking* insects and some terrestrial crustaceans are not subject to drift and can use steady winds directly as a compass to steer and maintain extended straight paths.

Fish and aquatic amphibians have an important receptor system, the lateral line, that responds to motion and its changes in the surrounding water. The lateral line of fish has sometimes been referred to as a distant touch sense; it occurs in many patterns in various kinds of fish and helps them school and helps blind cave fish avoid obstacles. Its receptor organs, the *neuromasts,* contain clusters of hair cells like those in the fish utriculus and semicircular canals. In many species the receptors lie in canals that may be elaborately branched along each side of the head and body; neuromasts may also be individually located on the surface of skin or scales or in pits. They respond to low frequency vibration and to water flow over the body surface or within the pits. Pressure differences or particle movements due to activity or currents in the water around the fish and in the canals cause excitation (or inhibition). Other nearby animals may activate the system perhaps from ranges of only 10 to 20 centimeters as in a fish school—or even the fish's own activity may do

A lateral line of neuromasts extends along each side of the body in such fish species as the *Monodactylus argenteus* (*below left*). The lateral line, which responds to low frequency vibrations and to water flow, is visible as a thin arched line. (*Below right*) In the herring and its relatives, the typical lateral line configuration is not found. Rather than running along the fish's sides, neuromasts are concentrated in its head. The red dots denote neuromasts found in the canals of a branched system. Others are scattered over the skin of the cheeks, just behind the mouth, and on body scales. Nevertheless, the sensor unit is still called the lateral line.

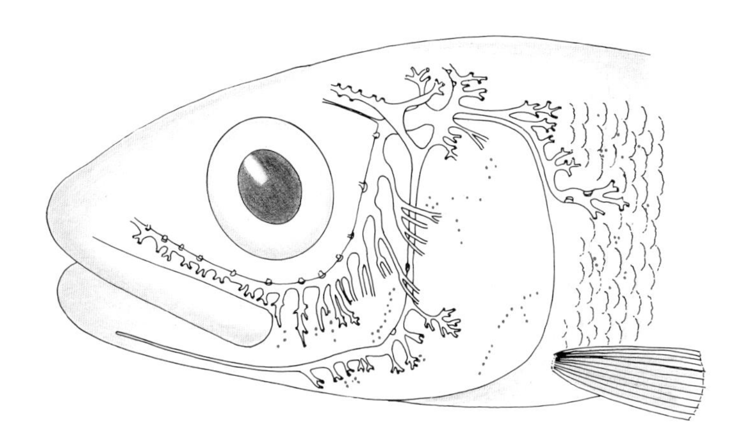

so. However, self-stimulation is apparently directly reduced or blocked by a matching inhibitory reflex from the central nervous system before and during the animal's own movement. The resulting stimulus pattern among the neuromasts signals the direction and nature of the disturbance so perceived. A mechanoreceptor system in the long antennae of certain pelagic shrimp has recently been found to respond remarkably like fish lateral lines.

Vibration sensing and hearing

Some animals can sense directions from vibrations in water or air—that is, they can localize the source of vibrations. Water striders and certain beetles that live on the surface of water can readily find prey or courting mates from the surface waves made by these other animals. The way whirligig beetles chase each other at the water surface seems remarkably like the behavior of house flies considered in Chapter 5. The beetles, however, do not use vision in their tracking but rather sense mechanical motion. They can do this because they have receptors at the base of their antennae, that are highly sensitive to waves on the water. Similarly the lateral lines of surface feeding fish and clawed frogs tell them the direction and distance to prey, predators, and rivals. Spiders are well known to be quickly alerted to the presence and direction of struggling prey by vibrations in their webs. Certain terrestrial mammals, reptiles, crabs, and especially frogs respond sensitively to vibrations transmitted by the ground on which they are standing.

With the help of hair cells and certain other sensory elements, many animals can detect regular repeated vibrations in air and water. These *displacement receptors* sense the movements of molecules and fine suspended matter. Because such displacements are transmitted only for short distances, these receptors are limited to short ranges. In contrast *pressure receptors*—the receptors typically responsible for *hearing*—can detect sound over much greater ranges. Pressure waves are transmitted much farther through both air and water than are vibrations. A humpback whale underwater may hear the song of another humpback from a distance as great as 165

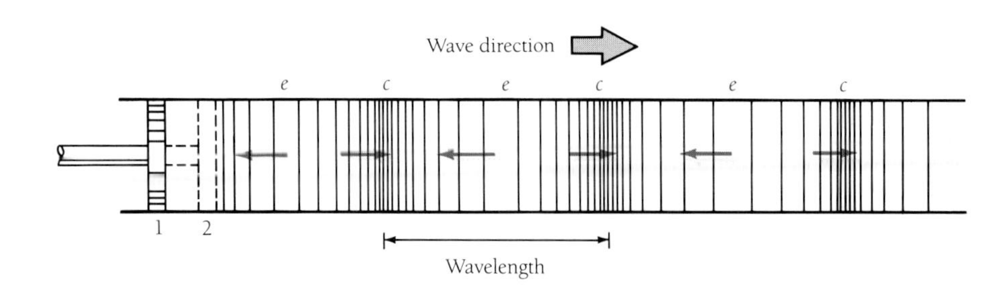

The piston oscillating between points 1 and 2 in this open air-filled tube models sound pressure waves. Moving right, the piston compresses nearby air (*c*), starting a pressure wave that moves rapidly in the same direction. The air expands as the piston pulls out (*e*). Although the distance traveled by individual air molecules is very short, the resulting pressure waves can travel long distances in the atmosphere or under water.

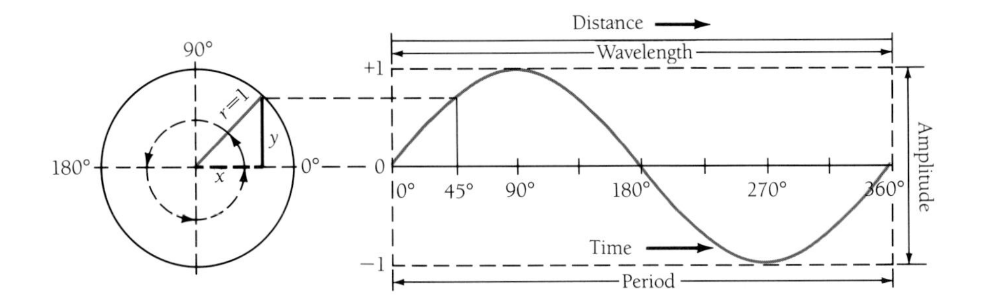

A vertical component y is always associated with radius r as it sweeps around a circle. A sine wave—the simplest periodic function—shows how y changes with time as r moves with a fixed speed. The wave has momentary values in time, varies smoothly through the cycle, and repeats. At 0° and 180°, its value is 0; at 90°, it reaches its maximum (+1 in the example here); and at 270°, it drops to its minimum (−1 here). The distance between those two points (2 in our example) is the amplitude of the wave. Its period is the time required (in seconds) for the completion of one cycle. The cycle also provides some additional values: frequency (the number of cycles per second) and wavelength (the distance covered per cycle). Multiplying the wave's frequency by its wavelength gives its velocity.

kilometers. Vibrations or waves that repeat periodically have three basic dimensions. *Frequency,* the number of oscillations repeated per unit time, is generally measured in cycles per second, or hertz. This largely determines what in music is called pitch. The amplitude of a vibration or wave, which is equivalent to height in a water wave, is primarily responsible for *loudness* in human hearing. Wave *shapes* vary widely from the simplest smooth form (a sine wave), to a fast spike, or a square wave. Wave shape is responsible for the "color" or *timbre* that makes an oboe sound different from a flute. Vibration sensing, hearing, and vision all depend on wave phenomena. Yet water and sound waves differ strongly from light waves in such important ways as speed of transmission, direction of vibration, wavelength, and interaction with sensors.

Humans hear sounds in the range from about 10 hertz to 20,000 hertz. We call sounds of lower frequency than this *infrasounds* and those of higher frequency *ultrasounds.* Wind, storms, and surf produce low frequency sounds that travel thousands of kilometers through the atmosphere. Homing pigeons, some large flightless birds, and guinea fowl detect these infrasounds and may use them for navigation. The Atlantic cod *Gadus morhua* also perceives infrasound that is present at high levels underwater. Actually the acoustics of air and water are quite different. Such loud underwater sounds as the songs of whales may be carried to great distances in the sea. Both the medium of transmission and the pressure waves' frequency affect the distances to which such waves travel before dying out. The transfer of airborne sounds into water and of underwater sounds into the air-filled middle ear of a whale, for instance, involve great losses of signal strength.

Hearing usually involves the perception of sound pressure waves in an elastic medium such as air or water. Unlike waves of light, which we have seen vibrate perpendicularly to their direction of travel, sound waves oscillate back and forth in the direction of their forward movement—or to say it another way, they compress and expand longitudinally. Sound waves can move a thin membrane, such as our eardrum. Its minute movements are carried mechanically through the middle ear to another membrane that contains thousands of receptive hair cells. Resultant waves in this basilar membrane stimulate the pattern of hair cells responsible for our perceiving particular sounds. Comparable hearing organs have evolved in many animals. For example, bush crickets have sound pressure receptors just below the knees in their front legs. Two slits below each knee open onto drumlike membranes that together respond to acoustic pressure gradients so that an internal receptor membrane on which hair cells lie is displaced accordingly.

Active and passive sensing

Directional sound signals perceived by an animal may come from nonbiological sources, other animals, or from the animal itself. A nonbiological source such as a waterfall or a lightning bolt provides strong directional signals that can be heard by appropriate ears. Prey, mates, or offspring may produce sounds that are orienting. Barn owls in the dark can accurately home in on the noise of a scurrying field mouse, for example, by using special, directionally keen ears evolved for this purpose. The calling song of the male cricket evokes a positive acoustic tracking response by receptive females: this behavior is essential to their reproduction. Infrasound produced by elephants may have similar reproductive functions. Any animal may itself actively produce particular sounds that are, like sonar pulses, reflected back to it with information about the direction, range, and nature of the reflector and whether it is fixed or moving. Bats and porpoises emit sounds of ultrasonic frequencies to find their way by echolocation. Acoustic signals emitted by one animal may evoke answering sounds from another animal of the same or a different species, thus establishing an auditory communication circuit. Such elaborations including mechanisms for producing as well as sensing signals is found not only for mechanoreceptors responsible for hearing but also for photoreceptors, chemoreceptors, and electroreceptors. When the same animal has both a generator and receiver for the same stimuli, the sensing is called *active*. *Passive* sensing is done by an animal having only a receiver. Among possible variations on an acoustic direction finder, the minimum requirement is one or more auditory receptors, usually paired, directionally discriminative, and able to sense the necessary sound features: intensity, frequency, duration, and also timbre if wave shape is to be recognized. Hearing organs with such capabilities have evolved repeatedly, being particularly striking in many insect types, in fish, frogs, toads, birds, and mammals. Not incidentally, these are also groups that have independently developed a wide range of sound-producing mechanisms important in vocal signaling, which culminates in human speech and language. Mating and social behavior in these various animals tend to be notably acoustical! Obviously direction finding remains a basic function in such active acoustic systems even when other important uses have been added.

Echolocation: bats

We may obtain a resonant introduction to acoustic direction finding by examining closely what is known about insect-eating bats hunting for nocturnal moths, which are their major prey. The story of the interactions of these bats and moths includes two types of ears, an ultrasonic pulse-generating mechanism, as well as bat search-and-pursuit behavior and insect evasive responses. Like the biosonar of porpoises, the echolocation of bats is mostly ultrasonic, being well above the human auditory range. The bats' high-frequency, short-wavelength sounds reflect well, carrying back precise details about obstacles to be avoided as well as flying insects. Having such an echolocation system for hunting allows bats to avoid daytime competition with the faster flying insect-eating birds. It does so by permitting such bats to maneuver skillfully in the dark without needing to see.

In echolocating, a bat emits the sound to be reflected as bursts of ultrasonic clicks or as sustained cries. The clicks are rapidly repeated; the cries may be constant frequency tones or frequency-modulated sounds—that is, sounds whose frequency changes with time. Such clicks and cries vary in different habitats, in particular phases of pursuit, and from species to species. The clicks produced are only a few milliseconds in duration and may be repeated during close pursuit at a surprising rate of 200 per second. The sustained constant frequency and frequency-modulated cries last 10 to 50 milliseconds or longer. Bats of many species emit sounds through their mouths, which act rather like a megaphone; bats of other species emit sounds through the nostrils and special associated structures.

Sounds produced by bats are as intense as the noise of nearby thunder or of a propeller aircraft at close range—loud enough to deafen the bat's own ears. Yet the

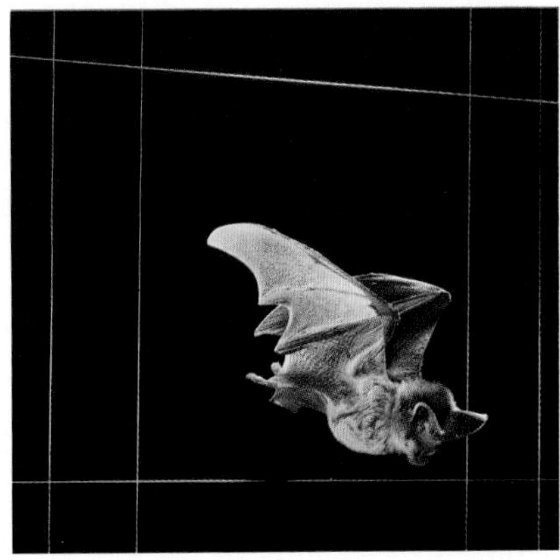

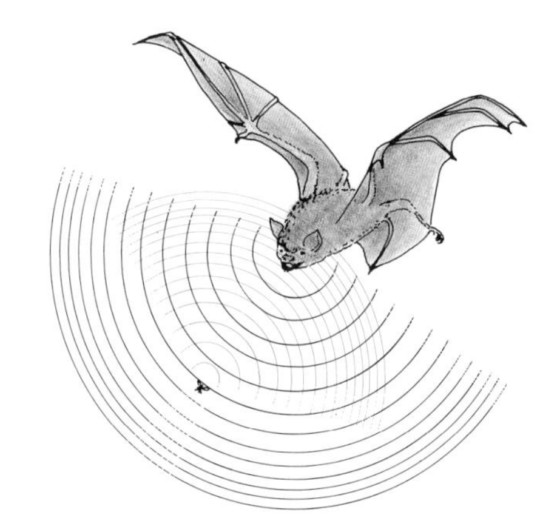

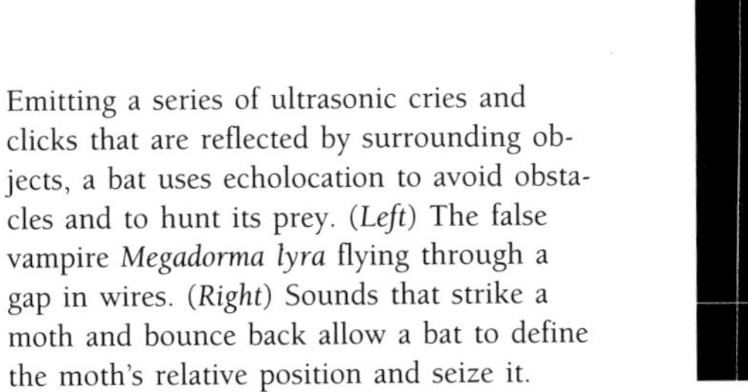

Emitting a series of ultrasonic cries and clicks that are reflected by surrounding objects, a bat uses echolocation to avoid obstacles and to hunt its prey. (*Left*) The false vampire *Megadorma lyra* flying through a gap in wires. (*Right*) Sounds that strike a moth and bounce back allow a bat to define the moth's relative position and seize it.

Like those of other bats, the receiver elements in the *Rhinolophus yunanensis'* echolocation system are shielded by the animal's own cries. The elements are turned down momentarily by muscles in the middle ears and other protective mechanisms.

animal needs an acute sense of hearing to catch returning echoes far fainter than its emitted sounds. As in a radar set, the receiver elements of the echolocation system are protected from the bat's own cries by being briefly turned far down by middle ear muscles and other mechanisms.

Any object hit by a bat's acoustic pulses will reflect them if it is large enough and not a perfect sound absorber, but an echo weakens rapidly with the distance between the animal and the reflector. If the resulting pressure waves of an echo reaching a bat's eardrums are above a threshold, the animal senses the direction and distance of the reflecting object. In certain instances a bat also detects speed and direction of a target's movement relative to its own, as well as its size, shape, and texture. Wingbeat frequencies of a flying moth may mark a returning sound with a telltale tremolo. The receiving end of the echolocation system extracts such detailed information from the acoustic stimulus patterns.

The paired internal ears of bats have a number of features that are particularly adapted for echolocation. One of these is high sensitivity to the main ultrasonic frequencies of the bat's own emitted cries. To detect target distance accurately, bat auditory systems must evaluate precisely the delay between the emitted cry and its echoed return. Because sound travels in air at about 342 meters a second, a distance of 1 meter is equivalent to less than 3 milliseconds in time. Hence, in using echolocation to decide whether a moth being pursued is 1 or 2 meters away, a bat has to identify very brief time delays. A bat determines the bearing of an echo's source by comparing loudness and relative delay perceived by its two ears; it also changes its flight direction and turns its head to gather more information about the target's location. Precise discriminations are made possible not only by adaptations in the bat's ears but also by a number of special kinds of central neurons in the brain. For instance in the mustached bat *Pteronotus parnellii* one special cortical analyzer consists of many neurons each responding only to an echo separated by a specific time delay from its originating cry and together constituting a graded sequence covering a range of delays from short to long. This bat's auditory cortex systematically classifies echoes from objects at various distances.

In bats with prominent constant frequency cries a substantive part of the auditory cortex is devoted to detecting and analyzing the frequency modulation of the echoes. If it is steady, an apparent difference in frequency between signal and reflection represents relative motion between bat and target. This is the Doppler effect, in which frequency seems to increase as two objects approach one another and to decrease as they move apart. If the frequency change varies rapidly, it can identify the wingbeats of insect prey.

A bat's acoustic world has a rather short range. The larger the object, the greater the range over which a bat can detect it. A sizable landmark might be detected at 100 to 200 meters. Yet for minute prey such as small insects, the bat's effective range shrinks to a few meters. Remember, even so, that bats are in effect "seeing in the

Katydids often alter their mating songs to avoid echolocating predators. Some communicate using vibrations that a bat's tracking system cannot discern.

dark." It is quite clear that they use echolocation at night in orienting to their local environment and in hunting. Whether echolocation helps them in their longer range movements is not certain.

Insect countermeasures

Some insects being pursued as prey detect the bats at much greater distances than the bats can echolocate them! Of course, the insect hears the very loud cries beamed specifically in its direction by the hunter while the bat listens for far weaker reflections from the small flying target. However, many nocturnal moths have evolved rather specialized bat-detecting ears. The remarkable ears of the night-flying moth *Prodenia eridania* seem specialized solely to help it escape from echolocating bats. These paired receptors are located on the sides of the moth's body. A slightly recessed eardrum covering an air sac is free to be vibrated by sound waves. Connected to the inside of this eardum are two receptor cells excited by its vibrations. They apparently provide the entire sensory output of this structure. Both respond over a range of 8 to 100 kilohertz with their peak sensitivity between 20 and 50 kilohertz, which corresponds well to the main hunting frequencies of the bats that are their predators. The two cells differ from one another, however, in being activated at different stimulus intensities. One reacts to relatively low sound levels and evokes evasive flight behavior in the moth. If the sound level nevertheless increases, the second sensor begins to respond. This induces the insect to stop flying and dive for the ground where it may be at least temporarily safe.

Various species of moths have rather similar ears with one to four (or more) receptor cells adapted to detect echolocating predators. These are remarkably specialized evolutionary responses of the prey to the bat's tracking system. Note that these antagonistic adaptations in prey and predator are like the development of weaponry (or Star Wars) by rival industrial powers. A nuclear missile on one side stimulates the invention of an antimissile missile by the rival. The latter clearly challenges the use of an anti-antimissile, and so on. Thus the so-called whispering bats have lowered their echolocating voices, making them less readily detectable by insects. Some katydids that could formerly be located by their mating songs have foiled their predators by changing these songs, communicating now by vibrations not detectable by bats. On the other hand some moths themselves, when closely chased, produce clicking sounds that apparently turn away pursuing bats, perhaps by startling them, or warning them of distastefulness, or jamming their ultrasonar.

Underwater echolocation

Echolocation is also known in a few birds and in some mammals other than bats, particularly in cetaceans (dolphins and other toothed whales). These marine mammals, like bats, generate and project ahead narrow beams of sound. Clicks of extremely short duration and as rapid as 200 per second, are common in a broad high-frequency range extending from 20 kilohertz to as high as 200 kilohertz. Whistles and other more complex sounds are also produced mainly to serve social needs. The extraordinary and well-known songs of humpback whales are not related to echolocation but rather may serve for communication like bird songs.

Although underwater acoustics differ in several important respects from those in air, the basic principles of cetacean sonar closely resemble those of the bat sonar. These aquatic species avoid obstacles and locate and identify prey at greater dis-

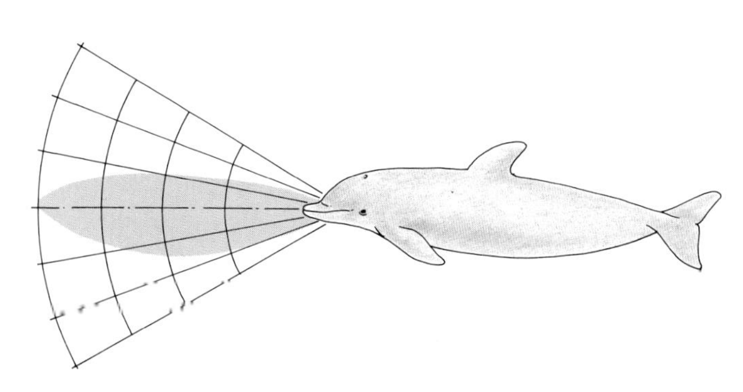

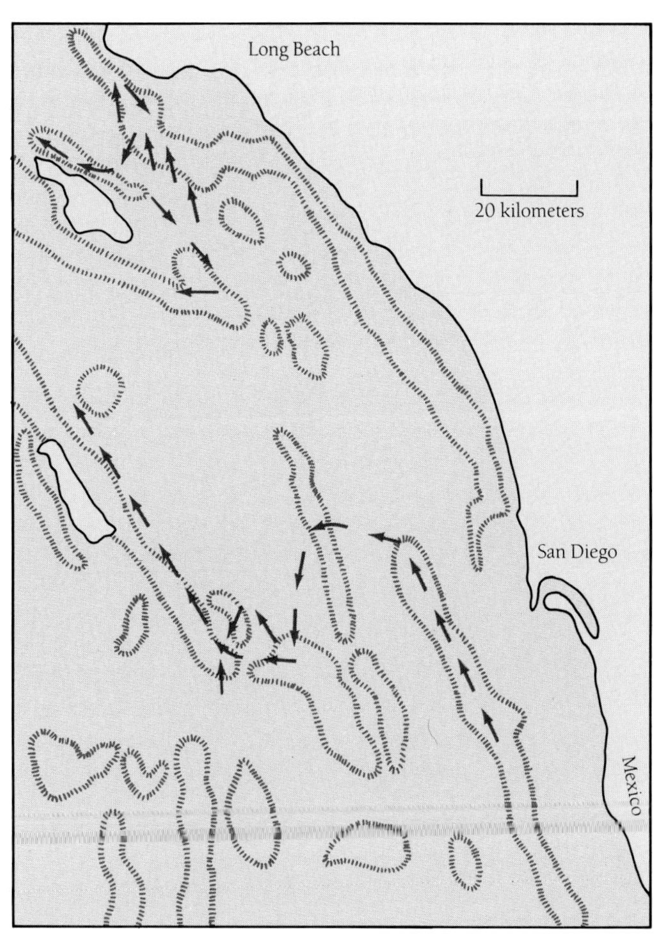

— Dolphin tracks

Underwater escarpments

Dolphins use echolocation to find food, avoid obstacles, and maybe even to navigate over longer distances. (*Left*) Repeated sound bursts—having a dominant frequency of 100 kilohertz—emanate from the head in a 20 to 30° beam. These directional high intensity clicks are reflected back by objects not completely absorbing them. Such echoes, which give a detailed sound picture of the environment at ranges exceeding 300 meters, are markedly better than the best underwater visibility. (*Right*) Tracking by radio transmitters attached to dolphins suggests that they follow landmarks revealed only by their echolocating systems. The series of arrows shown here are dolphin tracks that seem to follow submarine ridges and escarpments.

tances than bats can. The fact that free-swimming porpoises sometimes follow sea-bottom depth contours has suggested that their echolocation system may be used as a sonic depth finder to aid their navigation. However, biologists have not yet determined what the organ of sound emission and the detailed hearing mechanisms are. Most likely the concave skull and so-called melon in the forehead act to focus and project the outgoing sound pulses in the porpoise *Tursiops truncatus*. Part of the lower jaw may function in detecting and transferring the return echoes to the inner ear. This large aquatic mammal has proved considerably more cumbersome to study than bats, and much research remains to be done.

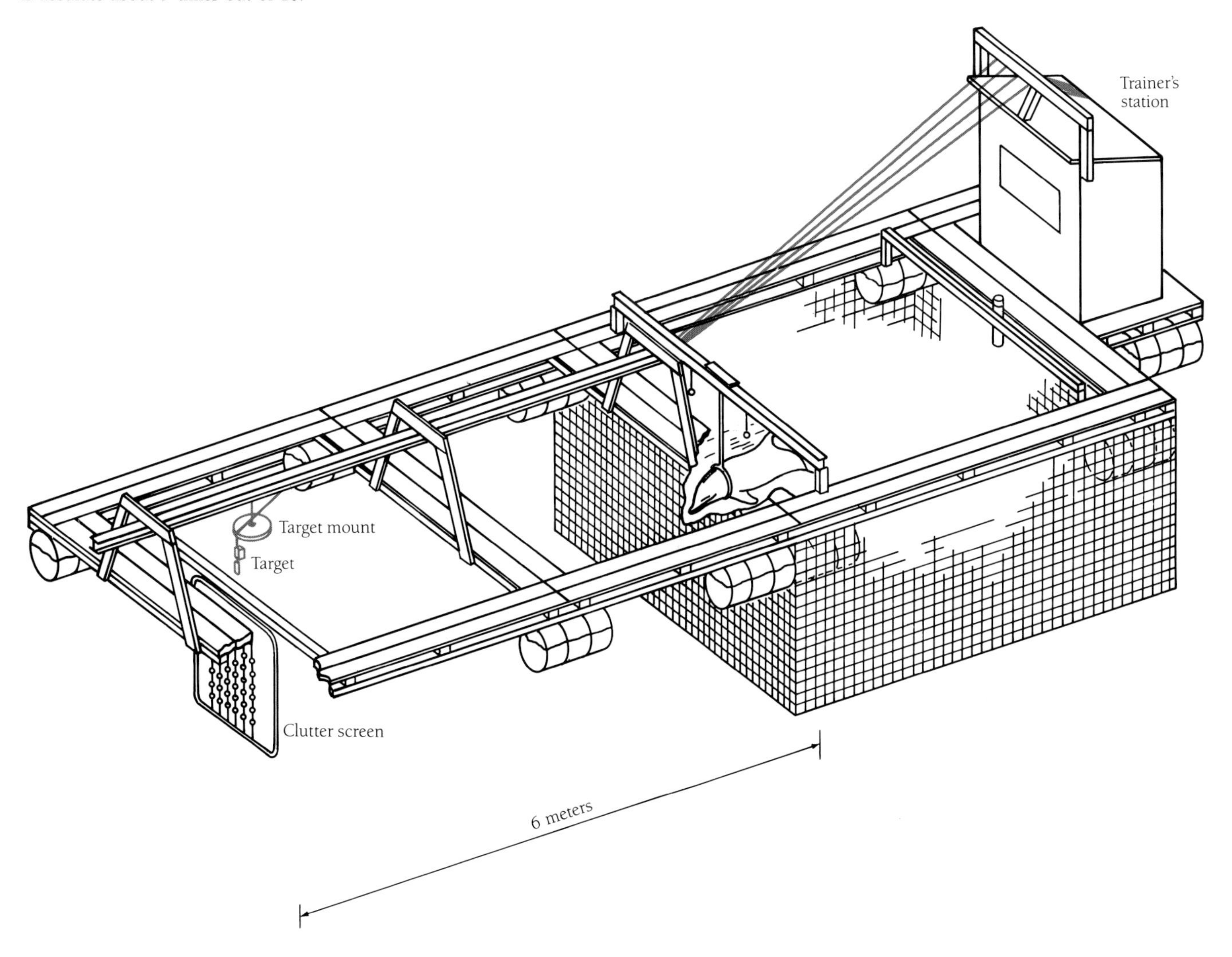

The precision of underwater echolocating systems has been gauged by experiments with animals penned in the sea. Swimming in a positioning loop, the dolphin here can distinguish between a target (a small metal cylinder) and potentially confusing background clutter (a screen holding 36 individual pieces of cork). When the clutter and target are at the same ranges, the dolphin's judgment is not as sharp—it makes errors roughly half the time—but when the screen is just 10 centimeters behind the cylinder, it is accurate about 9 times out of 10.

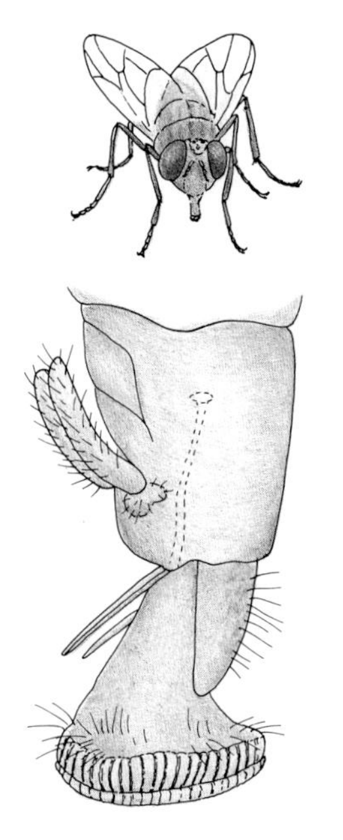

Contact chemoreceptors allow animals such as the fly to navigate over short distances. Located on its labellum, the insect's chemoreceptor cells respond to stimuli such as sugars and amino acids. The molecules thus serve as landmarks in the steering process. The labellum is seen here as the trunklike structure of the fly's head (*top*), in a side view (*middle*), and in a scanning electron micrograph of the ventral surface (*bottom*).

CHEMORECEPTIVE DIRECTION FINDING

When molecules and ions function as landmarks or occur as trails or gradients, they may provide valuable steering information to animals that perceive them. Like other receptor systems, smell, taste, and related chemical senses operate through changes in the resistance of special membranes. A particular stimulus may either excite or inhibit a cell. Certain chemoreceptor cells respond only to one particular type of molecule but others in contrast respond to a broad class of substances. The sense organs involved are usually rather small and inconspicuous.

Contact chemoreceptors like taste buds on our tongue or hairs on a fly's labellum (its taster) generally detect ions and molecules in solution. Such sensors are widely important in short-range direction finding related to feeding where localized amino acids, sugars, and other small organic molecules are common stimuli. In aquatic animals tasting often extends beyond the mouth and tongue as when it occurs on the barbels of a catfish and on other appendages like a lobster's claws and antennae as well as on the surface of the head or body generally. When water currents and gradients are present as they are in estuaries, taste receptors sensing salinities or various other elements dissolved in the water may provide orienting stimuli attractive or repellent to the organism. In air parallel functions usually involve organs of smell like the nasal chemoreceptors of mammals or the antennae of insects. These are sensitive to *volatile* airborne substances. Yet in fish, amino acids and other specific *dissolved* substances are also important stimuli of their olfactory system.

Chemical stimuli that animals respond to derive from many sources. Minerals and salts are ordinarily of nonbiological origin. Small organic molecules carrying the characteristic odors or tastes of plants and animals are, of course, of biological origin as are molecules specifically produced by organisms to be defensive or attractive such as the irritant sprays of certain beetles or the scent of skunks and flowers. Chemicals secreted by an animal as signals to other members of its species are called *pheromones* and function as sex attractants, for activating certain behaviors, for trail marking, and for individual identification. Kin and individual recognition are obviously highly important in social animals. Rodents can even be trained to discriminate among individuals of their own species by the odors of their urine; only identical twins are genetically similar enough to produce urine that smells the same to these sharp rodent noses.

Chemical direction finding in general seems to be one or another kind of trail following. Bloodhounds can follow for long distances traces of scent marking the passage of another animal. Mucus trails laid down by many intertidal marine mollusks, including chitons, limpets, and snails, both lubricate their creeping locomotion and serve to mark paths between their subhabitats. The "landmarks" tracked are ions or molecules that the navigating animal knows or has learned lead to a particular objective

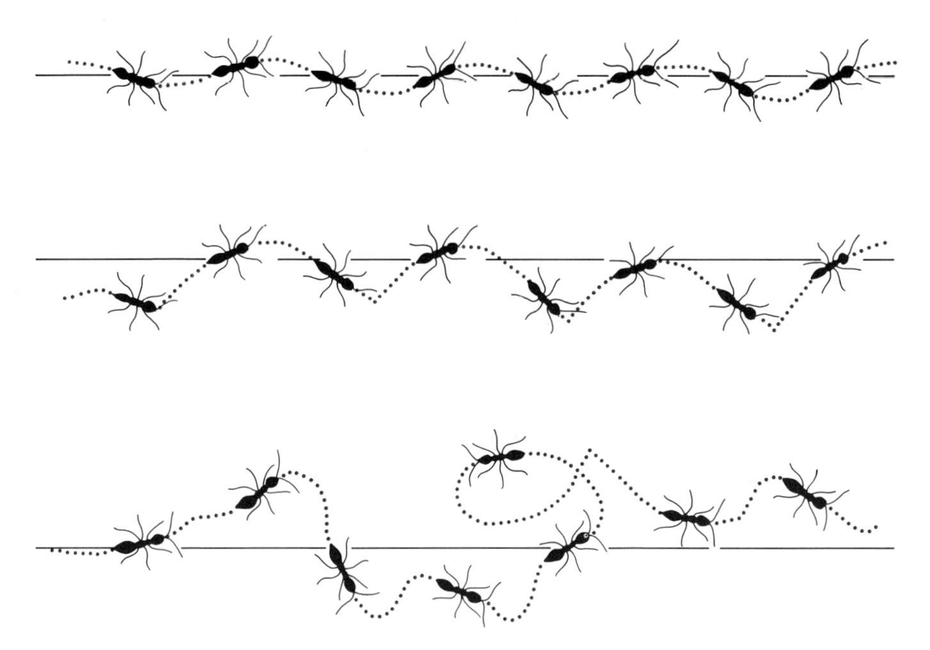

For social activities such as food gathering, ants use pheromone trails that serve as markers for recruits and nest mates. (*Left*) Walking normally, the North American desert ant *Novomersor* holds up its abdomen. Should it find prey too large to retrieve alone, however, the ant repeatedly lowers its abdomen and drops on the ground a trail of poison gland secretion to direct assistance. (*Right*) Chemoreception through their antennae allows ants to follow the odor trail. The diagram's solid lines represent the trails. A normal ant (shown at top) stays close to the pheromone markers. Despite the loss of an antenna, the middle ant still follows the trail. The bottom ant, with its right antenna bent to the left and its left antenna bent right, follows as well, although its course is rather crooked.

Insect pheromones

Ants deliberately use pheromones to mark trails between their nests and food sources to be followed by the many individuals in a colony. As the labeling molecules evaporate readily and disappear within a few minutes, the ants repeatedly renew the marking for as long as the paths are useful. Fresh trails of this kind are often polarized to indicate the goal's direction. Typically such molecular clues are recognizable only by members of the same species and some of them only by members of the same colony. So far we have mentioned only path markers laid down on the ground. Some animals leave chemical signals in air or water around them. Such information may permit quite spectacular moderate- to long-range feats of navigation. Although ions and molecules diffuse in such media, they do so slowly, except over short distances of a few millimeters or so. However, the atmosphere is characterized by winds and natural waters by currents that may distribute pheromones over great distances.

Male silk moths *Bombyx mori* follow airborne scents when searching for mates. A receptive female of this species stays in one place while emitting her sex-attractant pheromone, which diffuses or is carried downwind as a plume of signal molecules. A copulation-ready male in flight intercepts this aerial message. His large feathery antennae bear enormous numbers of unique chemoreceptors specifically adapted for this one job. Here again is a sensory unit with a single particular function, like the ears of the nocturnal moths that respond specifically to the echolocating cries of

In trying to locate a receptive female, the male silk moth uses a chemical compass. (*Left*) The male's feathery antennae, which bear chemoreceptive hair, are devoted solely to the reproductive task. (*Right*) Although females have no receptors of this kind, they have a pair of special glands at the abdomen tip that secrete the vaporous molecules that males can sense. Homing in on a zigzag course, a flying male usually turns to follow an odor plume downwind from the stationary female.

pursuing bats. In the chemical case, the male moth, as soon as he detects the pheromone molecules, changes his flight direction to turn into the wind. Then if he reaches the edge of the plume, he turns sharply back following a zigzag course. By maintaining an overall flight path both against the wind and back and forth across the axis of the plume, the male homes in on the signaling female from distances as great as several kilometers. Experiments have shown that the upwind trend of the male moth's tracking also depends on his visually perceiving landmarks in the environment. The zigzaging presumably increases his chances of reaching the target even if the pheromone is released irregularly or the wind shifts. This mechanism for following a scent has been widely studied not only for its inherent interest but also because it characterizes such highly destructive insect pests as gypsy moths and armyworm moths. Searching across the wind coupled with following odors upwind is widespread not only in various insect groups but also in other animals as well. Odor plumes are similarly followed in water too: Sharks use this behavior for finding prey and salmon for locating their home stream and breeding area.

About half of the 15,000 to 20,000 hairs on the antennae of a male silk moth are specialized for sensing the pheromone secreted by the female, a single molecule of which can trigger an impulse in a male's sensory neuron! A male responds behaviorally when his antennae have collected about 200 of the molecules. Biologists could measure the sensory and behavioral responses to one or a few molecule because the female silk moth's sex attractant had been chemically identified: This potent molecule produced by the species *Bombyx mori* is called bombykol and is an alcohol (a 16-carbon polyunsaturated hydrocarbon chain). The pioneer studies on this intriguing chemical compass were followed by a massive expansion of research on insect pheromones. Practical interest focused particularly on the potential importance of

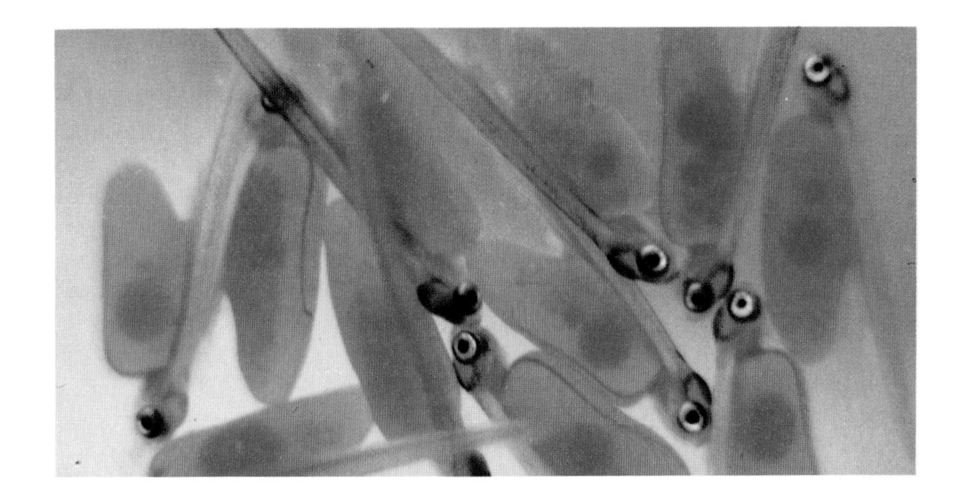

The return of breeding salmon to their natal streams shows the accuracy of chemical navigation. Although there is much research still to be done on the topic, some scientists believe that the natural substances dissolved in these home waters, sensed by the fish when they are first hatched, are remembered by salmon and serve to guide them when spawning. This notion is based on the idea that the fingerlings (such as those shown here) can be imprinted with a locally characteristic odor mixture.

these compounds for controlling pest species. More than 600 different insects are now known to utilize such chemical mating signals; About a quarter of these are moths. Each species has its own particular chemical mixture. As many as six or seven distinct molecular components present in certain ratios give high specificity to each natural pheromone.

Salmon homing

Another impressive feat of chemical navigation is the precise upriver return of breeding salmon to their natal streams described in Chapter 2. This behavior depends on substances carried in fresh water and perceived through the fish's sense of smell. Despite intense interest in this topic, no one yet knows for sure whether the odorous substances involved are pheromones or some other chemical markers.

An early decisive experiment proved that blocking the nasal cavities of migrating coho salmon with absorbent cotton upset their ability to choose the correct forks as they ascended their home river. Coho salmon (302 in all) were trapped soon after they had made a choice at a branching point in Issaquah Creek about 25 kilometers east of Seattle, Washington. The experimenters marked each fish to indicate whether it had been caught in the east fork or the west fork and plugged the nasal cavities of half the fish from each fork. The other fish in each sample—the controls—were not treated in any way except for the marking. The researchers carried all the coho down stream below the fork and released them. Traps were reset in each fork, and 143 of the salmon were recaptured. All controls retrapped in the west fork and 70 percent of the controls retrapped in the east fork had chosen the same fork as they did the first time. The salmon that were unable to smell, however, were recaptured in each branch in just those proportions expected if they were choosing at random.

What chemical clue are salmon following as they find their way up their home streams? One idea proposed in the 1950s by Arthur Hasler and Warren Wisby of the University of Wisconsin is that the critical molecules concerned are natural substances dissolved in the water and first sensed by the young fish in the area where they hatched. These odors—unique to the home stream—must be remembered through the intervening period of growth and maturation at sea, and serve to guide migrating adults back into the mouth of their native river, up stream, and home. According to this interpretation, molecules from soil run-off and from various terrestrial and aquatic plants identify the home water. For a brief period the fingerlings can be *imprinted* with such a locally characteristic odor mixture which they remember thereafter. By tracking this aroma 1 to 5 years later the returning fish can navigate precisely to its origin. A fair test of this rather challenging notion might involve the following steps. First, salmon hatchlings would be experimentally exposed, during the sensitive period for learning, to a dilute solution of some chemical not naturally present. They would then be released to go downstream and mature at sea. When it was time for these fish to return, a different stream could be baited with the imprinted substance. The explanation outlined would be supported if the fish were lured to a previously unvisited place by the learned clue.

Thirty years after it was proposed, Hasler and his colleagues carried out such an experiment in Lake Michigan where coho salmon have been introduced. An existing hatchery and stocking program supported this research, which involved the chemical imprinting and release of thousands of young salmon monitored over a 3-year period. Two groups of smolts were briefly exposed for imprinting in the hatchery to one or another of two odorous chemicals, morpholine and phenethyl alcohol, added in minute amounts. A third, control, group was not exposed to either chemical. This was a massive study with 5000 fish in each group for the first year and 10,000 in each group for the second. All fish, which had been fin-clipped to identify their group, were released near the northwest shore of the lake, halfway between two streams later to be labeled with the conditioning chemicals. Eighteen months later when these fish, now mature, were ready to travel upstream to breed, phenethyl alcohol was dripped into the Twin Rivers and morpholine into the Little Manitowoc River, 9 kilometers away. These two streams and seventeen others extending over a 200-kilometer neighboring stretch of shore were intensively monitored for marked fish. Although only about 3 percent of the 45,000 tagged fish were recaptured, they gave convincing evidence that salmon do smell their way home. Of all *recaptures,* 75 percent were imprinted individuals lured into the stream having "their" chemical. More than 93 percent of the morpholine-exposed salmon that were retaken, were caught in the Little Manitowoc River. Similarly, 92 percent of the recaptured phenethyl-alcohol-exposed fish were caught in Twin Rivers. In contrast, the control fish, not chemically imprinted, were widely distributed, being caught in fifteen of the nineteen sites.

Problems of mechanism

The Lake Michigan experiment clearly shows that coho salmon can respond to imprinted odors 18 months later by swimming toward an upstream spawning habitat marked with the odors. What is the relevance of these findings about artificial chemical imprinting to salmons' natural migration? Do results obtained with such nonnatural imprinting substances as morpholine and phenethyl alcohol simulate those normally occurring in nature? Would other substances often present in stream water evoke similar behavior? We do know from experiments that some organic molecules common in fish skin mucus strongly stimulate their olfactory receptors. In addition other tests prove that fish can discriminate between familiar and unfamiliar natural waters. But such data have not revealed what sort of olfactory clues salmon use in their upstream migration.

Actually, we do not know whether the keys to this chemical navigation are ions and minerals, organic material from other organisms, or chemical substances recently produced by members of the same population of fish. Some biologists believe that pheromones may be critical in upstream homing. Alternatively, mucus or fecal material produced by later generations still living in the home stream could provide the directional clues needed by returning spawners. Stimuli of this kind could account for the high specificity homing needed and are consistent with recent widespread data on kin recognition by fish as well as by insects and mammals.

Whatever the specific molecules may be, salmon—and probably many other migratory fish—certainly can follow a chemical trail in the water. In the ocean sockeye salmon from the Fraser River can apparently identify water from this river at a distance as great as 300 kilometers from its mouth. More fish complete spawning runs and reproduce in rivers whose outflow plume in the ocean is large and coherent than in rivers having small, diffuse plumes. The behavioral mechanism of this homing may well be similar to the turn into the wind and zigzag course of a male moth homing on the female sex pheromone. People have observed salmon swimming back and forth across the odor plume at river forks where decisions must be made. If a fish chooses the wrong branch, it stops, and drifts downstream until it encounters home-stream odors again.

Pigeon homing

In recent years a rather unexpected chemical compass has been reported (and denied!) for the homing pigeon. Certain biologists found that pigeons whose sense of smell is nonfunctional are less successful at returning to their loft from unfamiliar release points than normal ones. Close to home this effect is minor but it increases sharply for distances of more than a few kilometers. Pigeons may rely on *olfactory navigation* based either on directions from which odors come (a scent compass) or on a maplike distribution of odor patches in the environment (an olfactory mosaic). Wind directions bearing different odors, concentration gradients, or smell sequences may provide compass or map information to a pigeon. However, the reality of this proposed chemical navigation has been protested or denied by other biologists who have been unable to duplicate the original experiments despite amiable and cooperative attempts.

For one thing, the molecular nature of the directional or place cues is quite unknown although charcoal filters are reported to remove them from the air. Also, olfactory deprivation seems to decrease the speed and accuracy of pigeon homing. These conflicts are difficult to resolve because the birds may in fact be using other homing clues than those an experimenter chooses and deliberately varies. Recent experiments show that subtle factors like loft location and detailed training procedures influence navigational behavior. Because olfactory centers in the bird's brain are small relative to those in some animals, it does seem surprising that smell could be important for pigeon behavior. However, certain birds such as the kiwi, albatrosses, and vultures are indeed known to orient strongly to odors at least over short ranges. Olfactory orientation has repeatedly been evoked to explain homing at dusk or in the dark by burrow-nesting petrels, storm petrels, and shearwaters. Such birds apparently also use odor cues to locate floating oily food at sea.

THERMORECEPTIVE DIRECTION FINDING

Thermal stimuli work in many ways. Heat and cold can be *conducted* down a gradient in solids. They may also be carried by mass movements of air or water currents *(convection)*. Finally heat can be *radiated* through space as part of the spectrum of electromagnetic energy. Most of the time animals are either absorbing heat from their surroundings (and hence warming) or losing it (and hence cooling). Metabolism, which is universal in living things, ultimately releases heat. In birds and mammals metabolism maintains a warm fixed body temperature; most other animals dissipate heat, making their temperatures equal to or only slightly higher than those of their surroundings. These various relations between heat production, transfer, and dissipation establish the context in which animal thermoreceptors work. They also provide opportunities for these sense organs to function in direction finding. Probably the most spectacular of such applications is provided by infrared receptors that respond to electromagnetic energy at wavelengths slightly longer than the red we can see. It is as if animals with such sensors can "see" in the dark thermally even as bats "see" in the dark acoustically. Certain snakes actually detect infrared energy as it radiates through space.

Infrared reception

Visual pigments such as rhodopsin cannot function at wavelengths as long as those of infrared energy because the photons lack enough energy to trigger the first photo-

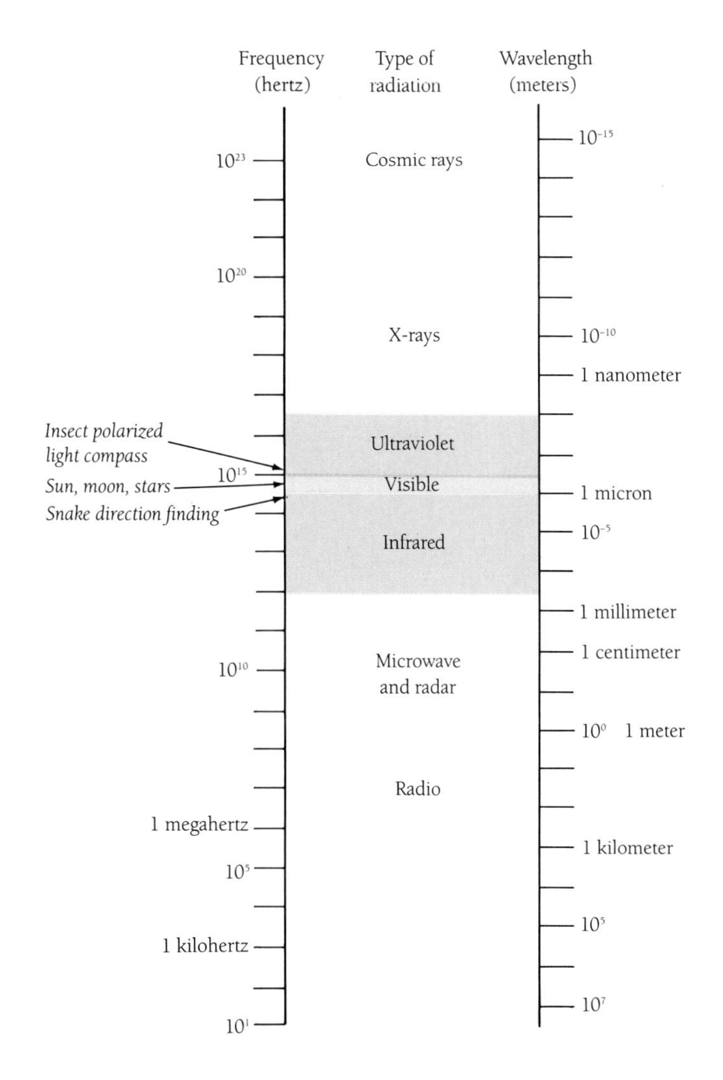

The electromagnetic energy spectrum spans the range from cosmic rays to the seemingly different long radio waves. Frequencies and wavelengths vary over a range from 10^{-15} to 10^{7} or, in decimals, from 0.000000000000001 to 10,000,000. Although, while navigating, animals use only a fraction of that spectrum, this portion is crucial to their direction finding. It centers on the narrowly restricted visible band at about 500 nanometers and extends into the near ultraviolet on the side of shorter wavelengths and into the near infrared toward the larger ones.

chemical step in vision. Photon energy decreases as wavelength increases. With wavelengths longer than about 750 nanometers, light has low probability of affecting the visual pigment. Heat radiation extends from about 750 to 1000 nanometers, where the "microwave" region begins. Some snakes and certain other animals that prey on warm-blooded animals localize prey with the help of infrared receptors, which are sensitive to wavelengths longer than those in the visible band. The best studied examples are the radiant heat organs on the faces of pit vipers and rattlesnakes. Using these sensors, the snakes detect prey over only short distances—generally 5 to 25 centimeters and perhaps occasionally up to half a meter. The

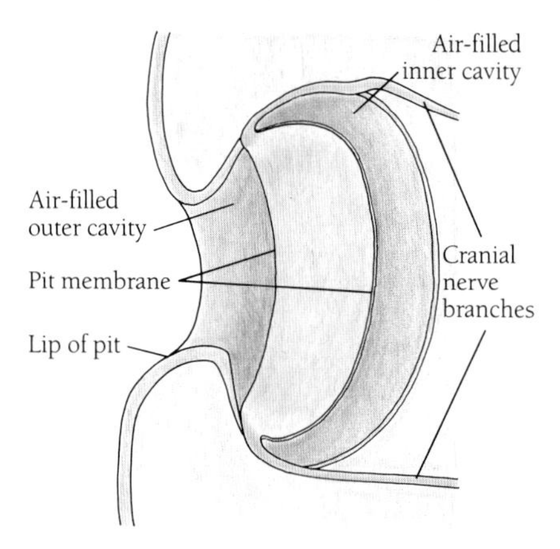

(*Top*) The golden eyelash viper *Bothrops schlegli* has radiant heat organs beneath and in front of the eyes. (*Bottom*) The receptive neurons of the rattlesnake pit lie in a deeply recessed membrane that is like the covering of a drum in being separated from the inner wall by an airfilled activity.

organs do, however, provide a sort of fuzzy, near-sighted "vision" for moving warm objects in rodent burrows or other dark places. The thermoreceptors project images onto the same brain areas that process visual maps from the eyes. Presumably, the snake's brain pools these independent sensory data to improve perception of the spatial environment. Elaborate processing of infrared information in the brain must account for the accuracy of the snake's strikes: The pit organs have wide angular fields—from 45 to 60°—yet blindfolded rattlesnakes strike within 5° or so of a small warm target such as a mouse. The thermoreceptors lie in a membrane richly innervated by thousands of finely branching nerve terminals. A temperature rise of only 0.003°C at the membrane changes the sensory response. The absorption of a minute amount of radiated heat is enough to warm the membrane and trigger the heat-sensitive mechanism.

Thermal preferences

Only a few animals can accurately sense directions from radiated heat, but many animals use their perception of temperature by other means for direction finding. Heat moves not only by radiation but also by conduction through solids and by circulation in gases and liquids. All animals can sense temperature through contact with the ground or with the surrounding air or water. Mobile animals given a choice of external temperatures usually locate themselves in areas having certain temperatures and avoid others. The temperature of choice varies from species to species and also within a species as conditions differ. Water near the freezing point is appropriate for arctic fish but lethal to tropical butterfly fish. The eggs of the brine shrimp *Artemia salina* stoutly resist freezing; adults, however, are easily killed by freezing. The global distribution of organisms is strongly correlated with worldwide thermal patterns and their daily, seasonal, and long-term changes. For instance, the developmental and seasonal migrations of bluefin tuna discussed in Chapter 2 can be seen as being associated—at least in bold strokes—with the changing distribution of water temperatures over the year. Such an association is clearly evident in the migration of shad. Birds and mammals that regulate their own body temperatures internally obviously have more freedom than other animals to live in environments of various temperatures. Their food plants or insect prey may, however, be more restricted by temperature and may limit the thermal range of the birds and mammals themselves.

Where temperature gradients, or warm or cold winds or currents, are present, directional information is available to any animal aware of them. Activating and orienting responses are often evoked by temperature differences, much as they are by the chemical stimuli already discussed. In many cases seasonal changes in temperature are undoubtedly a major factor in initiating migrations, but do they help animals find directions? Biologists have speculated about the extent to which temperature differences in space may be used by animals for guidance. For instance, relatively rapid ocean currents, or detached eddies and gyres of warm or cold water

The spatial accuracy of the rattlesnake's infrared system can be measured by presenting a warm object at various angles to the left or the right of the snake whose eyes are covered with blinders. Each circle represents a single strike of the snake and shows the angular error between the strike and the target. The average error of the strikes is less than 5 angular degrees.

often have rather steep thermal gradients along their margins. Can such gradients be used by migrating fish and whales as seamarks or underwater trails that could be followed with thermoreceptors? Oceanic migrators might derive a thermal compass or map from clues of this kind. After tracking European eels and Atlantic salmon underwater, one researcher concluded that these fish do respond to detailed temperature patterns in selecting their course. All animals certainly possess the sensors necessary for such behavior.

Shad migration

Migrations of western Atlantic populations of the shad *Alosa sapidissima* are closely associated with thermal cues. Like the salmon, the shad is an anadromous fish that matures in the sea for 3 to 6 years and then returns accurately to its native North American river to spawn. The freshwater breeding habitats of this species range from northern Florida to maritime Canada. In the fall juveniles descend from rivers and move out into ocean waters of the continental shelf. By winter they have converged from both north and south onto the mid Atlantic Bight north of Cape May and between 39 and 41° north latitude. There the species confines itself to a corridor having bottom temperatures between 3 and 15°C.

In the spring when ocean shelf waters become as warm as those of the home rivers, the various populations return to their breeding sites in a southwest to northeast sequence correlated with progressive warming along the coast. Thus in January shad reach the southernmost breeding grounds, which are in the Saint Johns River in Florida. They reach the northernmost ones, 2200 kilometers away in the Saint Lawrence River, as late as July. The species distribution and its migration are correlated to a remarkable degree with seawater and river temperatures. Yet it is not known whether thermal gradients and their seasonal changes can provide the needed directional information. Almost certainly thermal clues are supplemented in guiding shad to the home river by chemical characteristics of the parent stream, responses to currents in its freshwater outflow, and food availability. Some fishery biologists argue that a rather general sense of the direction home, expressed by populations, may be adequate to explain such return migrations. A corollary of this notion is that precise individual navigation may not be required. Future research should test the validity of these explanations for particular species and phases of the migratory cycle.

Hearing, smell, taste, and thermal sensitivity are sufficiently important human senses that we have a reasonably good intuitive feeling for their application to direction finding by animals. However, two additional senses—electrical and magnetic—covered in the next chapter are absent from or controversial in our sensory experiences. Yet they are certainly fascinating in their potential contributions to animal direction finding and compasses.

7

Electric and Magnetic Direction Finding

Electric and magnetic phenomena are widespread in nature and have in fact been known for a long time to be closely related physically. They are similar in establishing fields of force that have intensity and direction at any given point. Both kinds of fields are bipolar but the poles of electricity are positive and negative, those of magnetism, north and south. Static electric fields act on charged particles while fixed magnetic fields act on magnetic poles located within them. In both cases like units repel and unlike ones attract. Thus an electron, having a negative charge, is drawn toward the positive pole of an electric field, and the south pole of the magnetized needle in a compass points toward the earth's magnetic north pole. A moving electric charge generates a magnetic field around it and a moving magnetic pole establishes an electric field in its neighborhood. These various relations dictate the engineering of electric motors and generators as well as the possibilities of animals evolving receptors to sense such phenomena. Electric and magnetic fields are mutually involved in the radiation of light, heat, and other electromagnetic energy. Indeed the magnetic properties of a bar magnet or of the earth arise from selective orientation in the movement of electrons within their molecules. This fits with the fact that magnetic poles cannot be isolated but only occur as north and south pairs. Despite their tight interplay in the physical world, electrical phenomena seem far more important than magnetic ones in much of biology.

Receptors within these sharks detect minuscule electrical signals generated by the muscular activities of potential prey.

Bioelectricity—the electrical potentials and currents produced in cells and organisms—permits neurons, sense organs, muscles, and in fact all cell membranes to function. The voltages and current flow characterizing most biological activities are miniscule, but the electric organs of a few fish species provide an exception: These organs in the electric eel, for example, can produce shocks strong enough to knock

The sea slug *Tritonia* can orient itself to the earth's magnetic field. Sea slugs belong to the same group of mollusks as do snails but lack a shell.

down a horse standing nearby in shallow water. Many aquatic vertebrates have sense organs able to detect minute voltage differences, whether biological in origin or not. In contrast biomagnetism has been considered by many to be a subject about as worthy of serious twentieth-century scientific study as alchemy. Yet the important role of the magnetic compass in human navigation has stimulated speculation during the past 100 years or more about the importance of the earth's magnetic field for animal orientation and migration. Biologists have conducted many provocative experiments testing magnetic effects on organisms ranging from bacteria to humans. Particularly relevant to navigation are studies demonstrating to the satisfaction of many researchers that a wide variety of animals including insects and many vertebrates respond to magnetic fields as weak as those of the earth. However, other biologists disagree with this interpretation. Research attempting to identify magnetic receptors in a number of animals including insects and birds is underway, but much remains to be learned about sensory mechanisms and their use in animal navigation.

ELECTRIC DIRECTION FINDING

Electric phenomena depend on particles that determine or transfer charge. The charged particles in wires and transistors are electrons whereas those in living systems are ions, such as sodium and chloride. Whether such particles are positive or negative, together with their number, distribution, and flow determine how the system acts. A gradient of electric charges has voltage differences (E) between its two ends and between neighboring points along the way. If the charged particles are free to move, a current (I) will flow as long as the gradient persists. The rate at which current flows depends on the resistance (R) of the pathway for particle movement. Ohm's Law puts these basic relations together as $E = IR$. The resistance of a resting cell membrane is so high that little current flows through it despite the substantial resting potential between inside and out. On excitation, however, the gates open briefly and ions flow through. Voltages, currents, and resistance are also formed by larger scale phenomena of the world environment. There are, for example, both steady and transient small potentials in natural bodies of water. Steady voltage differences are geophysical in origin and are usually part of an electric field extending perhaps from one side of a pond to the other or from one region of an ocean current, such as the Gulf Stream, to a neighboring area. In contrast transient brief potentials underwater, which are like those of the alternating current of our commercial electric power, are usually (although not always) produced by animal activity such as electric organ discharges.

Many aquatic vertebrates ranging from lampreys to primitive mammals are now known to have sensors that detect and evaluate both sustained or fluctuating electric potentials underwater. Because a direct-current field has positive and negative poles like a battery and alternating-current waves come from a localized source, both may serve as direction indicators. Various navigational functions have been demonstrated or suggested for animal electroreceptors. Interestingly no such sensory mode has been found in terrestrial or aerial vertebrates or in any invertebrates.

Electric sense organs

Fresh water knife fish *Gymnarchus niloticus* served as subjects in pioneering behavioral experiments that demonstrated extraordinary electrical and magnetic sensitivity in aquatic vertebrates. These fish responded to a moving electric charge and to a small magnet moved rapidly along the side of the aquarium. Voltage differences as small as 0.03 microvolt per centimeter (1 microvolt = 0.000001 volt) in the water around the fish regularly evoked responses. Even though the voltage dropped at a rate equivalent to only 1 volt in more than 300 kilometers, there was no doubt that the animals detected this slight gradient. However, sense organs necessary for this remarkable electroreception were not identified until some years later when they were recognized by other biologists studying how electrical signals help sharks locate their prey.

Electroreception seems to have been an ancient development in evolution because it is found in lampreys, sharks, skates, and rays, all of which appeared early in

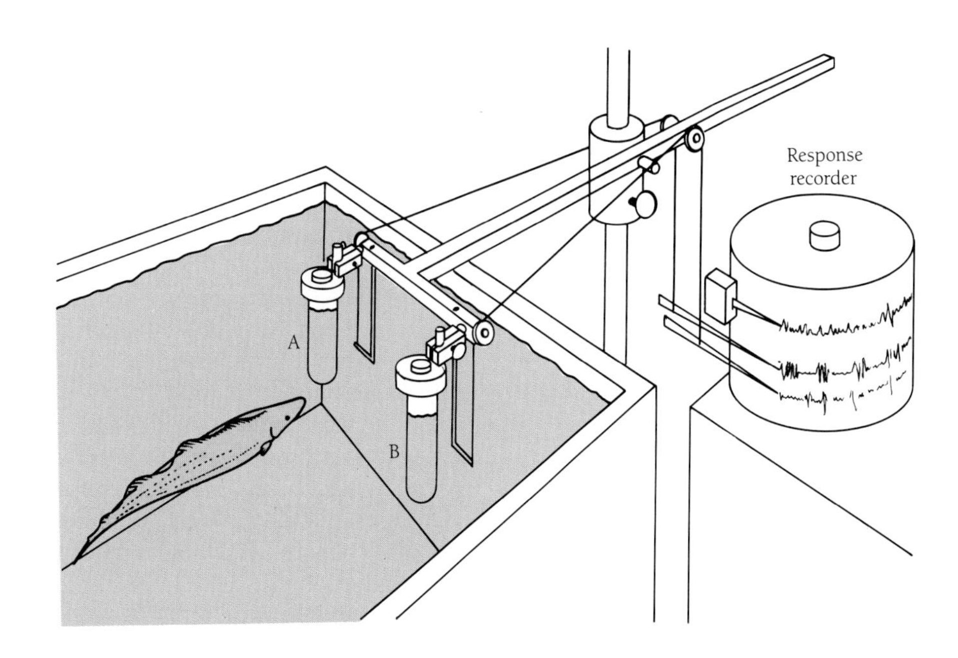

The knife fish was trained to locate food by using its amazing ability to discriminate between objects of different electrical conductivity. Two tubes, A and B, were placed in the aquarium with the fish. One tube was filled with tapwater, an electrical conductor, and the other with paraffin, a nonconductor. The contents of the tubes were switched frequently, but food was always hidden behind the tube containing the tapwater. The fish learned to feed behind the electrical conductor and to ignore the nonconducting tube. Subsequent experiments showed that the fish can discriminate between slight differences in conductivity, such as between tapwater and a mixture of tap and distilled water.

The duckbilled platypus, a primitive mammal, is among the wide range of aquatic vertebrates endowed with electroreceptors. Electroreceptors can sense minute differences in voltage reaching various parts of the animal's body surface.

vertebrate history. Sturgeons, lungfish, bichirs, paddlefish, and coelacanths—all surviving relics of ancient Paleozoic bony fishes living 5 to 6 hundred million years ago—also have electroreceptors. The recent demonstration of such sensors in two groups of amphibians, including newts, as well as in the duck-billed platypus, (an egg-laying Australian mammal), shows that electroreception evolved independently and sporadically among various kinds of aquatic vertebrates. Although teleosts (the bony fishes) have been subjects for many studies of this sense, only a few, widely separated subgroups of this large diverse group have electroreceptive species. So far no major migrators such as salmon, tuna, shad, and herring have been found to have electroreception. Nevertheless, the well-studied reactions of knife fish and the prowess of sharks, rays, and catfish in orienting by means of electroreceptors imply that such organs could be used for navigation by any animals endowed with them.

Receptors that sense voltage differences in water are of many types but they fall into two major classes, called tuberous and ampullary. All seem to function as voltmeters that assess external potential differences between various spots on the animal's body surface. Only electrolocating fish have tuberous organs; they also have ampullary organs, as do other fish and other aquatic vertebrates. An ampullary organ is highly responsive to low-frequency sustained voltage differences (ranging from about 0.1 hertz up to 10 hertz or so). The structures in sharks and rays that are

Two major types of electroreceptors. The ampullary organ is sensitive to steady, low-frequency voltage differences produced in the animal's environment. These electrical signals are conducted by the gel filling the neck and ampulla and stimulate the sensory (receptor) cells, which transmit the signal to the nerve. In contrast, the tuberous organs, which are just partially sunk in the skin, occur only in electrolocating fish, which have electric organs. The electric organs continually generate high-frequency pulses, which create an electric field around the animal. Objects near the fish distort the field; the distortion is "read" by the tuberous sensory system.

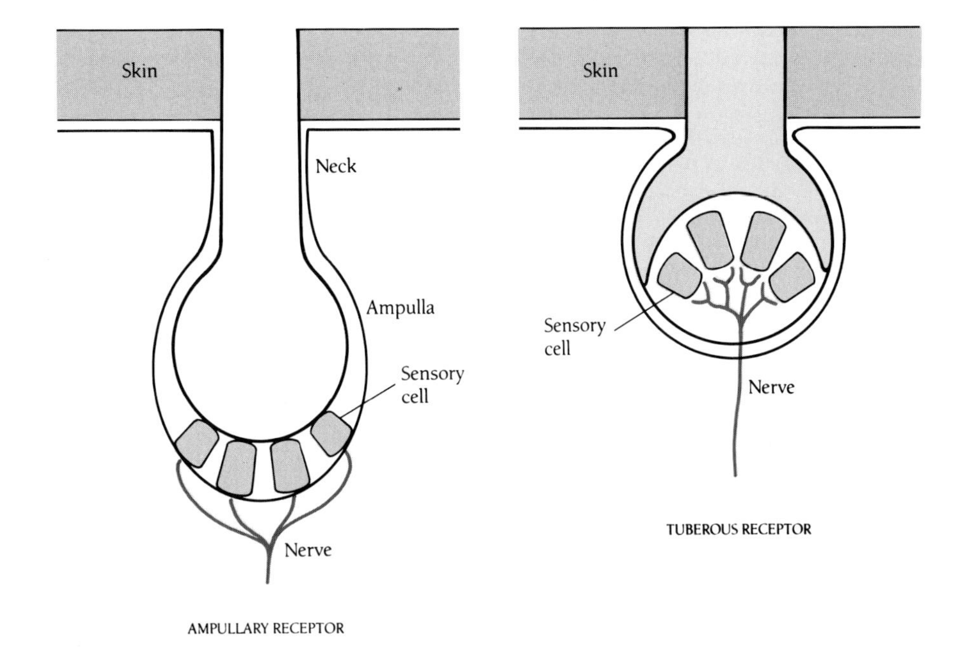

receptors, called the ampullae of Lorenzini, have long been known to comparative anatomists. Such ampullary organs may be thought of as direct-current electroreceptors. Located on and around the snout and elsewhere, their sensory cells lie in small bulbs (the ampullae) well below the skin surface, to which they are connected by long thin open-ended tubes. The walls of the tubes have remarkably high electrical resistance but their gel-filled cavities act as ideal electrical conductors, transmitting with little loss of voltage, potentials at the skin surface inward to the receptor cells themselves. The sensitivity thresholds of the organs are quite remarkable: less than 10 microvolts per centimeter in freshwater forms; as little as 0.005 microvolts per centimeter for marine sharks.

There is good evidence that natural voltage gradients, whether biological or geophysical in origin, give electroreceptive animals directional clues. The experimental data are extensive, particularly for freshwater catfish as well as for marine sharks and allied species. Although it was suggested early that animals in both these groups have electrical sensors, compelling demonstrations and explanations came only after sensitive and sophisticated electronic equipment was used to dispel erroneous ideas. The ampullae of Lorenzini were not so long ago still believed to be thermoreceptors or mechanoreceptors even though their structure and location both seem inappropriate for such functions.

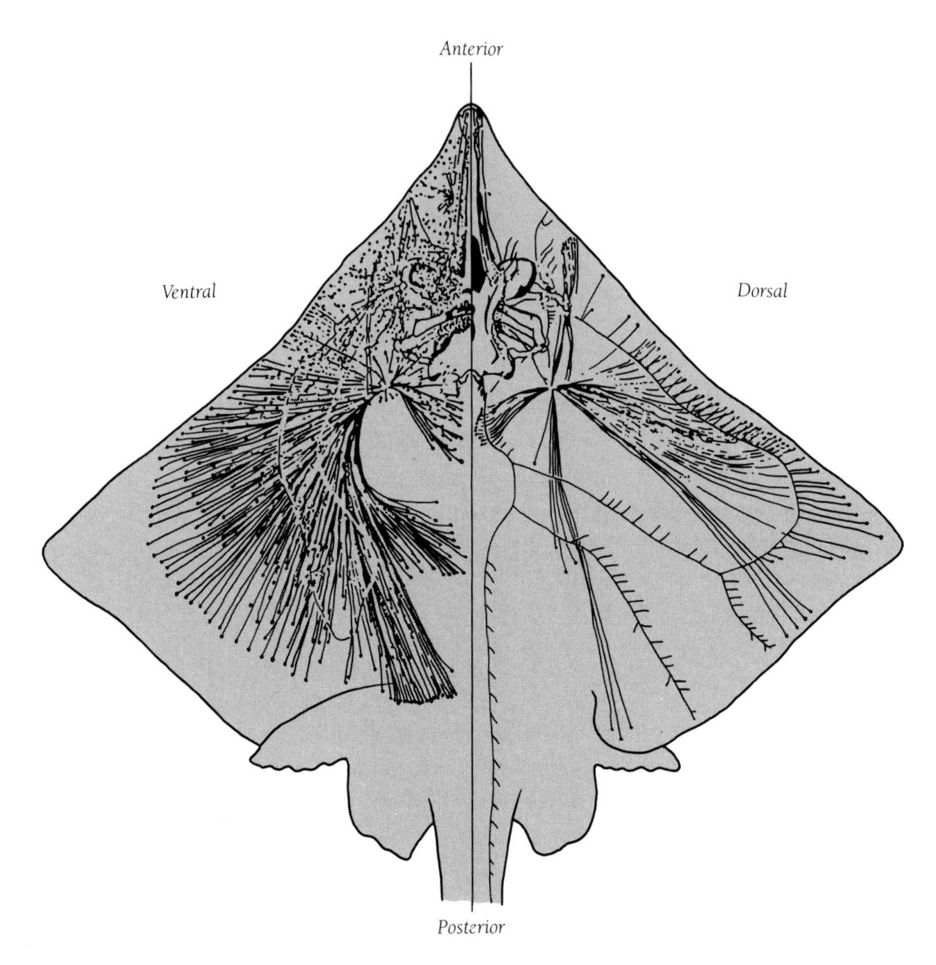

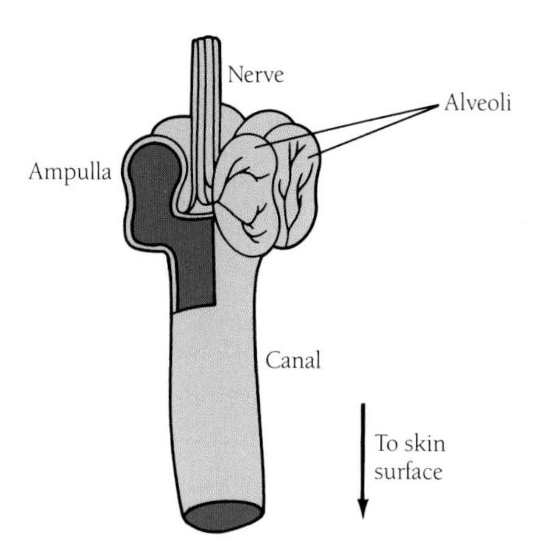

The electroreceptor system of sharks, skates, and rays is an elaborate array of tiny ampullary organs, connected to the skin by fine tubes. (*Right*) As seen in the drawing of the skate *Rais laevis,* some of the tubes are remarkably long. *R. laevis* has nearly 1700 ampullary organs (called ampullae of Lorenzini), 89 percent of which open onto the ventral surface, particularly around the mouth and snout. (*Below*) An ampulla consists of four to eleven grapelike alveoli lined with a layer of several thousand receptor cells. Each ampulla measures the voltage where its canal opens onto the skin relative to a uniform voltage within the fish. This intricate sensory mechanism allows the skate to locate live animal food and to orient itself to electric and magnetic fields.

Use in hunting prey

The first convincing evidence that they are in fact electroreceptors came from demonstrations of nerve impulses from snout ampullae in the dogfish *Scyliorhinus canicula*. Mechanical and thermal stimuli evoked only minor responses from these small sharks, whereas voltage differences applied externally resulted in massive neural activity. Somewhat earlier measurements had shown that voltage differences as small as 1 microvolt per centimeter stimulated the receptors in both a shark and a ray.

Demonstration of the direction-finding usefulness of electroreceptors began with the observation that a hungry dogfish can readily detect live flatfish in the sand. From ranges of 10 to 25 centimeters it attacks directly places where prey are buried, digs them out, and devours them. To investigate how a dogfish knows where to dig, an experimenter covered a flatfish with an inverted dish made of agar gel and buried them in the sand. The dogfish located these prey without difficulty. The agar blocked visual, chemical, and mechanical cues from reaching the shark, but did not screen out transient electrical changes generated by muscular activity in this living prey.

Results from three additional experiments reinforced the conclusion that a shark follows electrical signals to find living food. Nonliving chunks of fish under agar did not attract the dogfish. Covering a live flatfish with a thin electrically insulating sheet of plastic as well as with agar completely blocked the shark's detection of its prey. Finally a pair of metal electrodes buried in the sand to simulate the flatfish's bioelectric field evoked aggressive feeding responses from the dogfish. Free-swimming sharks in the sea have also attacked such active electrodes as if they were prey.

An electric compass

The freshwater catfish *Ictalurus nebulosus* is similarly attracted to buried electrically active electrodes as if to food. Furthermore this fish readily learns to use minute direct-current gradients in its environment to localize an object. Training experiments prove that the direction of a hiding place at the edge of the circular aquarium can be learned and repeatedly chosen if a weak direct-current electric field is maintained in the tank. For instance if the positive electric pole is to the north and the shelter to the east during training, the fish will search for the shelter to the south when the positive electric pole is shifted to the east. Hence the experimental electric

The catfish *Ictalurus melas* can use electrical gradients in its environment for direction finding.

field is being used as a reference to find the right course to the hiding place. This is in fact a close match to the way human navigators use a magnetic compass.

If further research shows that many aquatic animals rely on such direction finding, electroreception will be recognized as an important mechanism for underwater navigation. There are a number of geochemical and geophysical phenomena that produce electrical potentials greater than those known to be detectable by freshwater electroreceptive fish. Surprisingly large voltage gradients have been measured in freshwater ponds and streams, as great as 300 microvolts per centimeter over short distances. Boundaries or gradients of temperature, acidity-alkalinity, and other chemical or ionic content would be expected to produce such differences in charge distribution. Although, the actual sources of the potentials observed in nature and their possible usefulness to electroreceptive fish remain largely unknown, there is some evidence that local electric features of the seafloor can be used for piloting by species such as dogfish that feed on the ocean bottom. In addition substantial potential gradients exist in the seawater itself because of electric currents induced by water movements through the earth's magnetic field.

ELECTROLOCATION

All animals generate electrical potentials to polarize cell membranes and in the activity of their sense organs, nerves, and muscles. Also a number of fish such as the electric catfish, the electric eel, and the knife fish, have evolved organs that produce electric pulses. In a few cases the shocks are powerful enough to disable prey: the electric eel *Electrophorus electricus* can send out a pulse of several hundred volts. More commonly, much weaker discharges from electric organs help fish communicate and make them capable of *electrolocation,* used by the fish much as porpoises and bats use acoustic echolocation. Pulses of several hundred millivolts to several volts serve here in the fish. For electrolocation an animal must have both an organ that generates electric pulses and sensors on its body surface that detect the resulting field around it. Nearby objects of different conductivities distort the field pattern set up by the pulsing organs. The fish thus perceives an electric "image" of its close surroundings that changes as the fish or other objects move. An electrolocating fish constantly emits low-voltage pulses and, aided by its imaging system, deftly avoids obstacles, orients to the bottom, swims backwards through crevices, locates prey, and discriminates neatly between objects having only minute differences in electrical resistance. It is easy for such a fish to discriminate in the laboratory, for instance, between a rod made of glass, which is an insulator with high resistance, and one made of metal, which is a good conductor with low resistance. Indeed, so sensitive is the system that the fish can readily distinguish electrically a glass rod 4 millimeters in diameter from a 6 millimeter one. In action this discrimination endows the animal with a remarkable avenue of perception quite unknown to humans.

The electric organ of the knife fish generates pulses that set up an electric field for which the head acts as the positive pole and the tail as the negative. This field will be distorted by any object whose electrical resistance differs from water, such as the conductor (blue circle) and insulator (green circle) shown in the drawing. Notice that the conductor distorts the field by drawing together the lines of force nearest to it. The insulator distorts in the opposite direction by spreading apart the lines of force. These alterations in the electric field are monitored by the tuberous organs found in the fish's lateral lines.

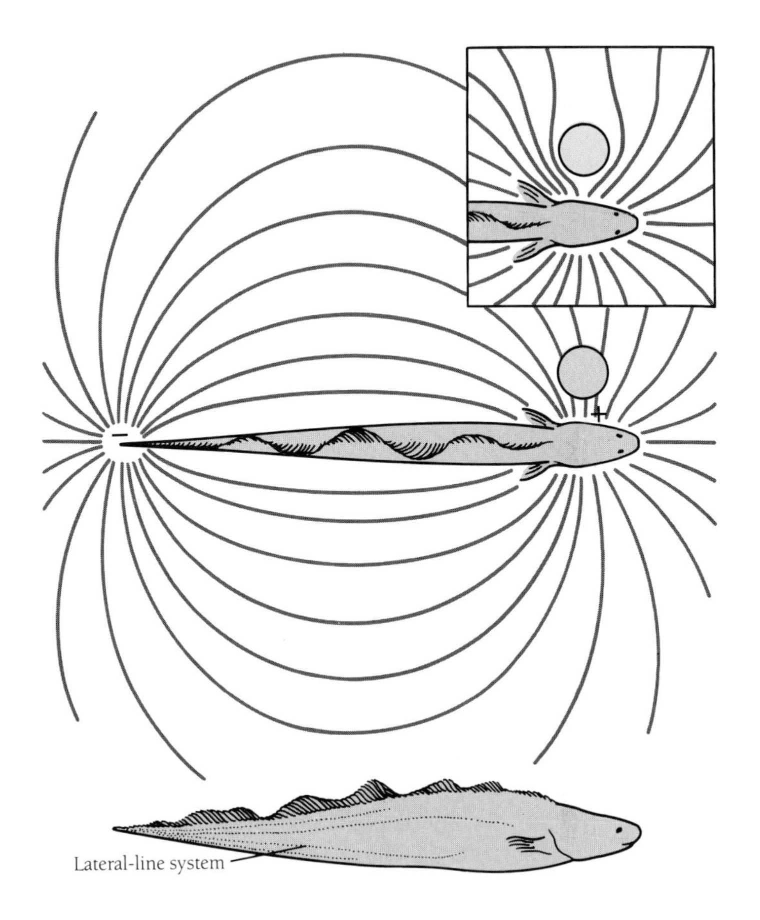

The electric field generated by such fish is perceived in detail by numerous tuberous organs on the surface of their heads and bodies that respond to rapid transient signals. These electroreceptors are particularly sensitive to the frequencies of the fish's own discharges. When there are no objects nearby, a standard symmetrical electric field is set up around the animal by its pulses. Differing from sound or light, the electric pulses and resulting field are not propagated but are static. As soon as something whose electrical resistance differs from that of the water enters the area, the standard electric pattern is distorted in a predictable manner. The animal's electric sense organs "read" this change. In this way, the fish localizes obstacles or prey from distances or a few centimeters up to perhaps 40 centimeters. Electrolocating fish can detect the output from another fish's electric organ output over greater distances—from a few tenths of a meter to a few meters. In responding to electric signals from an intruder of the same species, a fish defending its territory seems to locate the external source of electric pulses by swimming along the field's lines of force. In any case the static nature of the electric pulses limits the distance the pulses travel but at the same time quite fully preserves their precise waveform. Quite a number of freshwater fish in Africa and South America have electrolocation systems. They all live in murky water or are active at night. Possibly because the much lower resistance of seawater tends to short-circuit the buildup of electric fields, electrolocation seems to be restricted to freshwater animals.

Each species of electric fish emits continuous discharges at its own characteristic frequency in the range from 5 to 1800 hertz. Members of one species in the genus *Eigenmannia* have individual "private" signal rates somewhere between 200 and 400

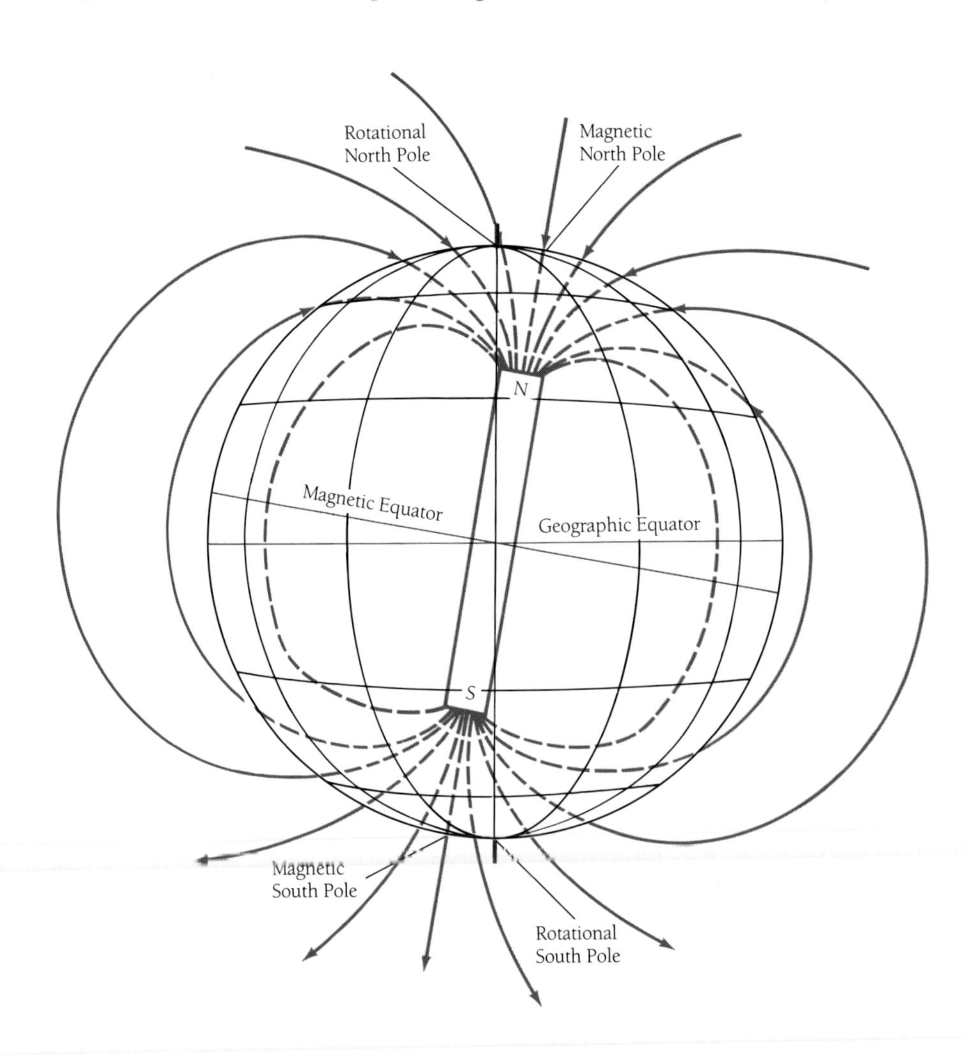

The earth acts like a huge bar magnet with north and south magnetic poles. For navigation, the resulting magnetic field at the earth's surface provides important information about global directions.

hertz. Although the frequency emitted by an individual or a species is often highly stable, frequencies may nevertheless vary between day and night, between the sexes, or with different seasons, hormone levels, or temperatures. Each species also has its own waveform of electric output. It varies in different cases from discrete voltage spikes or short pulses, cleanly separated from one another in time, to continuous waves of periodically varying voltage. Pulses produced by different species last from about 175 microseconds to 8 milliseconds. Hence any stream with many electric fish in it is rife with their signals, which are being used for localization, identification, and communication. If a number of species are together, the result may be a babel! In a single river system in West Africa, a biologist found more than twenty kinds of electric fishes, each generating its own characteristic electric pulses. The situation intensifies in the dry season when the increasingly restricted and turbid water may literally crawl with electric fish. What keeps them from jamming one another's electrolocation systems?

MAGNETIC DIRECTION FINDING

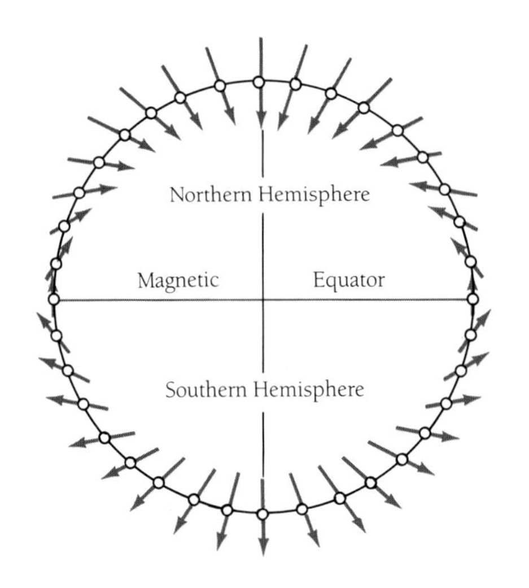

The polarity (arrow heads), dip (angles to the earth's surface), and intensity (length of arrows) of the magnetic field at the earth's surface vary systematically with latitude as shown here. Although a human navigator's magnetic compass measures only the field's polarity in a horizontal plane, animal compasses might sense, in addition, dip as it changes with latitude, the direction of maximum dip (a measure for north or south), or total intensity, which becomes stronger with increasing latitude.

There are two natural sources of magnetic force that conceivably might influence animal behavior, including navigation. One of these is the rather minor localized influence of lodestone (magnetic iron oxide or magnetite) and other naturally occurring magnetic materials. The other—and by far the major—source is the field generated by the earth, which acts like an enormous bar magnet. Magnetic lines of force diverge widely from the earth's magnetic poles emerging upward around the south pole and entering downward around the north pole. These lines of force, which collectively are the earth's magnetic field, cause the south pole of a freely suspended magnet to incline downward in the northern hemisphere, to be horizontal relative to the surface of the earth near the equator, and to incline upward in the southern hemisphere. Hence the magnetic field has two components, one horizontal and one vertical, at any given point on earth. The horizontal component establishes a north-south series of magnetic meridians. At the magnetic equator equidistant from the magnetic poles, the lines of force are horizontal and the vertical component near zero. Over the magnetic poles the horizontal component is near zero and the lines of force almost vertical. As we will see, the vertical component, which is called the *magnetic dip,* seems to be important for some animals' navigation.

Because the magnetic poles are displaced from the geographic rotational poles, there is usually a difference between the true north-south meridian and magnetic north. This *magnetic variation*, which may be large in some places and times, changes usually slowly over a period of years as the magnetic poles drift. In addition to its horizontal and vertical *directional* components, the geomagnetic field also has a third dimension, *intensity* (or strength). This increases with latitude from 30,000 nanoTeslas at the magnetic equator to less than 70,000 nanoTeslas at the poles. (A nanoTesla

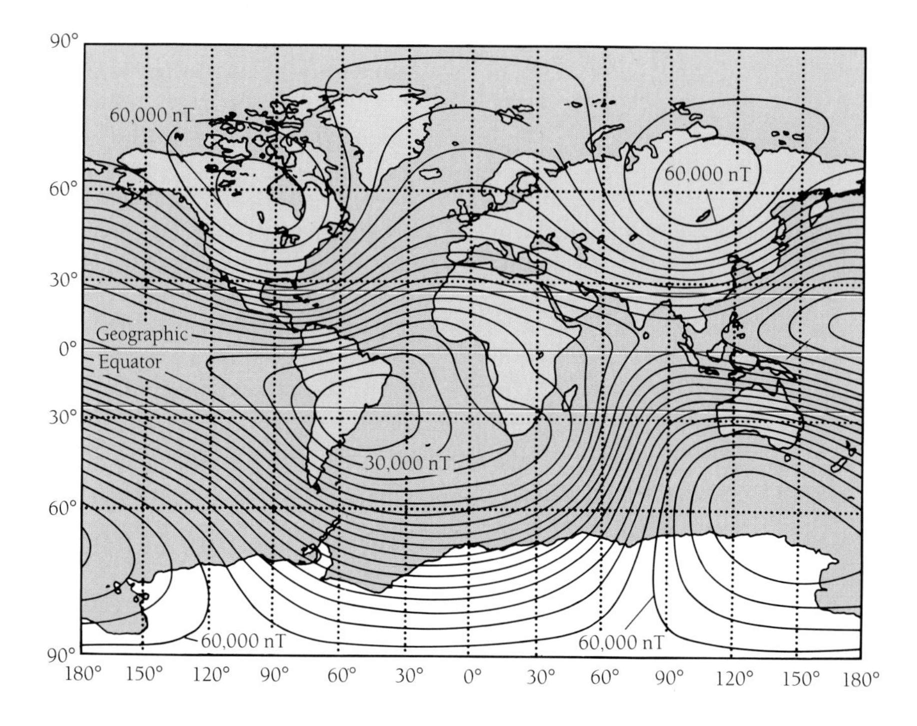

As this world map of geomagnetic intensity shows, variations in the intensity of the earth's magnetic field make more complex patterns than the diagram on page 155 suggests. Nevertheless, a rather simple pattern of wavy east-west contours covers much of the world. If an animal navigator could sense gradual increases or decreases in magnetic intensity, it would have at least a crude compass to use between the southern United States and Central America, the Caribbean and northern South America, or up and down the Australian and the east Asian coasts.

is a unit of magnetic intensity.) The field intensity also decreases, as might be expected, with distance from the earth's surface. These natural magnetic forces are all very weak compared to those in an electric motor, for instance, or even in a toy magnet. Indeed this intensity is so low that many biologists have doubted that animals could possibly perceive the earth's magnetic field.

The earth's field changes its detailed features in the course of a day, a year and over longer periods. The time scales on which changes in the earth's magnetism have been measured or estimated range from seconds to millions of years. From a navigational point of view long-range variations are relatively small over most animals' lifetimes. In the long run, however, for time intervals great enough to include the evolution of migratory behavior, such changes can be dramatic, extending to a complete reversal of the magnetic poles. This is known to have happened many times during the earth's history at intervals varying widely from 20,000 years to 70 million years. The geomagnetic field, as well as its long-range changes, are believed to originate from the slow circulation of molten elements in the earth's core.

In addition to these global features, there are nontrivial but low-amplitude, localized, magnetic irregularities both in space and time. The *spatial* irregularities reflect the uneven distribution of magnetic materials in the earth. Peculiar, site-specific behavior of homing pigeons has often been attributed to such geographic factors. If this is correct, these birds, like honey bees, must be able to discriminate subtle differences in field intensity. Migrating birds change their flight altitude—but not their course—over a strong localized magnetic anomaly in central Sweden. It seems that these migrators sense the steep changes in the earth's field in that area and respond to them. Recent detailed analysis of strandings by whales and dolphins suggest that they may be relying on magnetic piloting or course holding by following magnetic patterns laid down geologically on the seafloor. Following these "landmarks" too far could lead to beaching. Marked geomagnetic variations over *time* depend, among other things, on recurrent but variable solar storms that bombard the earth's field with strong radiation. Such variations, however, affect the overall strength only by a factor ranging from 0.01 to 10 percent. Even so, errors in the homing of pigeons, for instance, have been reported to be correlated with these magnetic fluctuations.

Magnetic effects on bee behavior

The idea that sensing natural magnetic fields helps animals steer their migration or homing has been a popular one for many years. Many biologists, however, could not take it seriously because behavioral and physiological data to support it were scanty. The topic began to seem respectable when some large-scale experiments on honey bees were reported in 1972. Earlier, von Frisch had discovered that communication dances performed on vertical honeycombs usually showed moderate errors (up to 15°) from their correct bearing. These errors varied in size and direction with time of day as well as with the compass alignment of the surface on which the bees were dancing. The new finding in 1972 showed that such mistakes in orientation are well correlated with small daily fluctuations in the intensity of the earth's magnetic field. There is a basic 24-hour periodicity in the fluctuations (and consequently in the errors) but this is overlaid by additional variations both regular and irregular.

Interestingly the bees make no errors when the dance directions on the vertical comb coincide with the direction of greatest dip in the local magnetic field. When the experimenters put large active Helmholtz coils, which alter or eliminate the earth's field, around the hive, it became clear that changes in the magnetic field of the dancing area consistently modified orientation errors. Indeed, when the magnetic field was 95 percent canceled out by the coils, the bees made no systematic errors! Note that such behavior does not suggest that the bees' perception of the earth's magnetic field is used for a compass. Instead it somehow modifies their translation of information about solar azimuth into corresponding responses to gravity. Nevertheless, these results provide dramatic evidence of the bees' remarkable sensitivity to environmental magnetic fields.

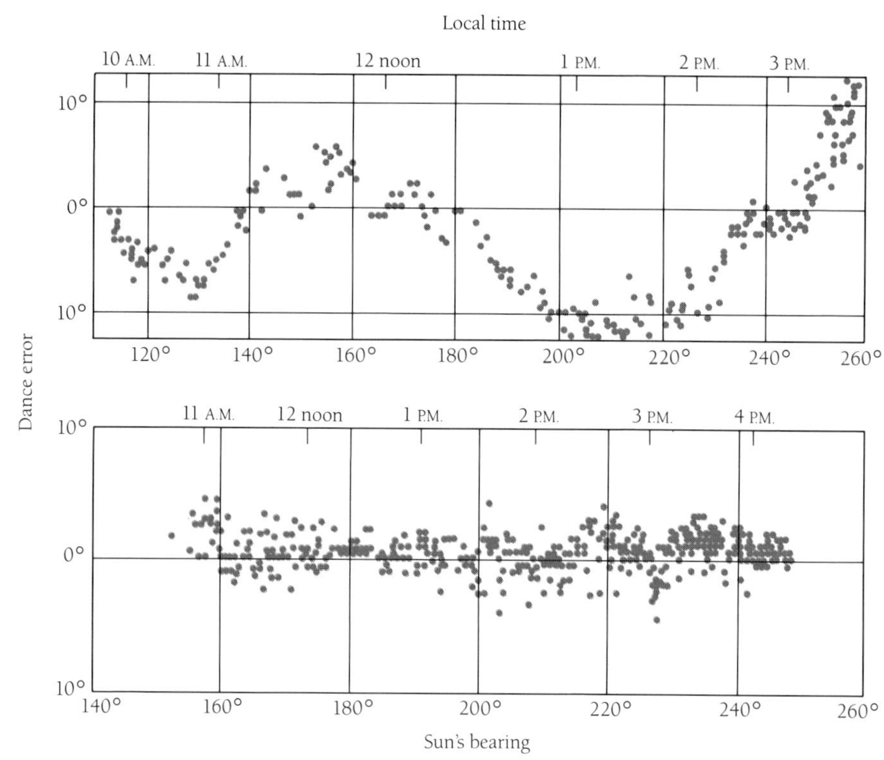

Daily fluctuations in the intensity of the earth's magnetic field influence the honey bee's compass orientation by 5 to 15°. (*Top*) Errors in dance directions vary systematically with time of day. (*Bottom*) When the magnetic field was experimentally reduced the errors were all less than 5° and showed no major trends.

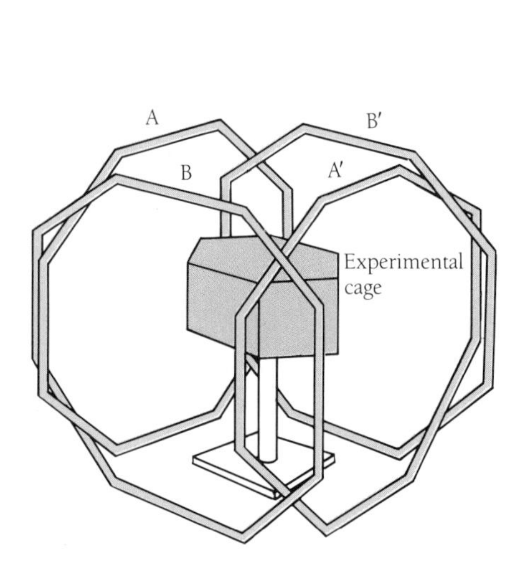

Experimental apparatus for studying the effect of the earth's magnetic field on animals. Electric current flowing in a coil generates a magnetic field around it. With the two pairs of Helmholtz coils, AA′ and BB′, the earth's magnetic field within the animal's cage can be intensified, weakened, or canceled altogether.

Later a real compasslike response to the earth's magnetic field was demonstrated for honey bees. When constructing new combs they collectively line them up accurately in a particular magnetically determined direction. Comparable magnetic alignment had been found in wasps and in the nests of certain termites. Another well-oriented behavior appears when honey bees perform on a horizontal comb surface under diffuse lights without a view of the sun or blue sky and begins only after 2 or 3 weeks of *disorientated* dancing! Then the figure-eight dances point sharply to the four main quarters of the compass and less strongly to the four 45° intermediate directions. The dances are not at all related, as they usually are, to the bearing and distance of food sources but are directed arbitrarily to magnetic north, northeast, east, southeast, south, southwest, west, and northwest. Hence the whole pattern looks uncannily like a stylized eight-point human compass card. How such stereotyped magnetic performances relate to practical direction finding is not known. If Helmholtz coils are used to cancel the earth's normal magnetic field, random dance orientation replaces this behavioral compass rose.

These extensive experiments on the effects of geomagnetism on honey bee behavior are provocative because they suggest that foraging bees can take local magnetism into account for piloting or for identifying places. However, further research is needed to discover how the bees perceive the earth's magnetic field and how they may use it in their navigation.

Magnetic responses in birds

Directional restlessness in caged migratory birds demonstrates that a number of species can maintain their normal free-flying seasonal flight headings in the absence of all known compass clues except the earth's magnetic field. Shielding the birds from that field or changing its intensity and orientation with Helmholtz coils supports this important conclusion. Such a magnetic compass was first demonstrated for the European robin in 1965. Sensing the poleward direction depends on the dip of the magnetic lines of force. Hence the compass is an inclination detector that does not distinguish north from south magnetic poles and could not function near the magnetic equator where there is no dip. Interestingly its usefulness would not be affected by pole reversal of the earth's magnetic field. The birds must be exposed for at least several days to a given magnetic intensity before they can use the earth's magnetic field effectively as a compass. A number of other old-world passerines as well as several new-world species like the indigo bunting and the bobolink, all nocturnal migrators, have a comparable magnetic compass.

Pigeons seem to depend on geomagnetic information to home successfully under certain conditions. In clear weather, they use information from the sun and sky for direction finding. But under *fully overcast skies*, small magnets fastened to their backs or tiny Helmholtz coils on top of the head disorient them. Low-intensity,

The European robin uses a magnetic compass to maintain the correct heading during migratory flights.

oscillating magnetic fields upset their initial orientation on release even under clear skies. From unfledged chicks to experienced adults, ring-billed gulls seem to use magnetic compasses. Both small magnets attached to their heads and intense magnetic storms appear to disorient their migration and homing. For this species the solar compass seems to be a critical partner for direction finding with the earth's magnetic field. Both are essential in choosing the correct bearing. Juvenile birds of other species seem to depend on geomagnetic forces as their first compass, which they later use to calibrate the celestial compasses, from which they take their bearing during migration. Experienced pigeons, for instance, no longer depend solely on their magnetic direction finding and indeed prefer to use the sun as a compass if it is visible. The explanation for such a two-tiered compass is not certain but no doubt relates to convenience and reliability. Moreover, recent behavioral work on salamanders and electrical recordings from the brain of the bobolink, an impressive transequatorial migrator, suggest that these animals have more than one magnetic-sensing system.

Magnetic effects in fishes

As we have seen, visual piloting and celestial navigation are of little use to aquatic species. How do long-range oceanic migrators such as salmon, tuna, and eels find their way? They are not known to have electroreceptors but biologists have wondered whether such fish perceive the earth's magnetic field directly. Despite considerable research on several salmon species, the evidence for a magnetic receptor is incomplete and sometimes contradictory. For instance, juvenile sockeye salmon can orient to weak magnetic fields, but they do not seem to have magnetite particles, which provide the mechanism for one well-understood biological compass. In contrast, appropriate chains of magnetite particles were found in anterior skull cartilage of the closely related chinook salmon, but the requisite behavioral experiments have not yet been done on this species.

Fry of still another Pacific salmon, the chum, demonstrated some *partial* responses to 90° rotation of the horizontal component of the earth's magnetic field, but the experimental work did not include a search for magnetite particles. By inserting magnetized wires into the skin of the fish's heads, biologists conducted an experiment similar to that in which magnets were attached to the heads of homing pigeons. Unlike the birds the experimental chum fry in a distorted magnetic field behaved no differently than the others being observed in the normal field. Thus current information on geomagnetic orientation in Pacific salmon is rather difficult to generalize. Similarly incomplete and sometimes conflicting reports of magnetic effects in the Atlantic eel make it difficult to prove their importance in navigation. Training experiments on young yellowfin tuna have shown that they can discriminate between a normal earth's magnetic field and an altered field but not much more than that is known.

Magnetic orientation in humans?

In addition to those animals already mentioned, the ability to sense magnetism has been attributed to protozoans, flatworms, snails, amphibians, reptiles, and mammals including humans. In 1980 a British biologist presented evidence that people, without any special experience or training, know the compass direction they are facing after they have been spun around while seated wearing a blindfold and earmuffs; people also, according to this same study, can point toward their starting location after making a circuitous bus trip while blindfolded. That these responses depend in part on sensing magnetism was shown by altering the earth's magnetic field around the observers' heads with magnets or electromagnetic coils inside helmets they were wearing. When this was done, part of the orienting ability was lost.

Within the next 4 or 5 years, a number of other researchers around the world attempted to repeat these provocative results. All of them concluded that their results did not support the idea that humans have the magnetic navigational ability deduced from the earlier data. In the face of such universal rejection, the original claims seemed dubious, or at best, anomalous. Recently, however, the first biologist challenged his numerous critics by reanalyzing as much of their total data as could reasonably be pooled. He concluded that quite contrary to the critics' analyses the combined results do in fact support the idea that humans even when deprived of visual cues can sense direction, noting that the probability is less than one in a thousand that the results were due to chance. The hypothesis that the human azimuth perception depends to some extent on sensing the earth's magnetic field is estimated to be statistically significant with the chance of only one in twenty that the result was due to random effects. Whether this reassessment will be accepted by those who failed to reach such conclusions and will be sustained by further research remain to be seen. At present the opposing conclusions seem to depend on different ways of applying statistics to the same data.

In this connection it is fair to say that not all of the claims that numerous other animals can sense magnetic fields have been convincing. As mentioned, some skeptics remain unconvinced by much of the data. Furthermore, the general relevance of magnetic sensitivity to *migratory* navigation remains to be directly proven *in the field*. It has been previously reported, as evidence for the influence of geomagnetism on bird navigation, that the degree to which migrating land birds tend to get off course at night corresponds to the existing level of magnetic noise or "storms." A recent review, however, failed to confirm this correspondence.

The question of magnetic receptors

Questions about the behavioral and migratory evidence supporting magnetic navigation by birds are not dismissed by what little we know so far about the needed sensory mechanisms. What sort of receptor for sensing magnetic fields could an animal have? Have receptor organs evolved for responding directly to one or more dimensions of global and local magnetic fields? To date experimenters have found the search for such sense organs in animals highly elusive; some scientists doubt that magnetic receptors exist. However, *electro*receptors of the sort described above could well provide an animal with the means for wresting directional and perhaps other navigational information from magnetic fields. Such employment of electric sensors is possible because movement of a charged particle in a magnetic field induces an electric field. This means that ions displaced in an animal by its own swimming through the earth's field could set up a voltage gradient detectable by electroreceptors. Because the detailed way in which magnetic and electric fields interact is too technical to analyze here, the following account is largely descriptive.

Electroreceptor responses to magnetic fields

The best-documented case for a magnetic compass involving a known sense organ in an animal is mediated by electroreceptors stimulated through an electromagnetic mechanism. The animal is the small sting ray *Urolophus halleri*. To test the hypothesis that these fish use a magnetic compass, weak magnetic fields were set up in their aquarium to simulate the earth's field in the fish's natural habitat. The rays were

The sting ray *Urolophus jamaicensis* learns to locate a feeding place by means of the ampullae of Lorenzini. Even more interesting is its ability to use the earth's magnetic field as a compass.

trained to feed in the earth's magnetic and geographic east. However, if the simulated field was then rotated, the fish sought food in the east as defined by that imposed shift. Presumably the fish detected the voltage gradient that arose from their own movement in the aquarium's magnetic field. Both direction and polarity sensed by their electroreceptors provided the necessary compass reference. Note that the fish receptor differs from that of warblers and sparrows by distinguishing magnetic north from south.

This experimental work did not test whether the ampullae of Lorenzini participated in the behavior, but some Soviet biologists studied the responses of single neurons from these receptors in two related species. The nerve impulses in these species, besides being sensitive to electrical stimuli, were excited or inhibited by weak magnetic fields provided either the water or the fish were moving or the stimulus was changing. As the water or the fish moved faster, the strength of these responses increased, as it did when the magnetic stimulus changed rapidly. Such dynamic effects would be expected from the proposed electromagnetic interaction. These results generally support the idea that sting rays and sharks use electroreceptors to provide a compass dependent on the local geomagnetic field.

Possible use in the sea

What do these experimental results on rays and skates tell us about how their electroreceptors can monitor the earth's magnetic field? The key point is that a voltage gradient, if only a very weak one, must be induced on the fish's skin by its movement through the earth's field or by an ocean current that is moving sufficiently fast. Substantial voltage gradients have been measured in the ocean. These are largely attributed to "motional electric" effects. They arise from water movements, like the Gulf Stream or tidal currents, flowing through the earth's magnetic field at high latitudes where the vertical component of this field is substantial. Around the magnetic equator, where the magnetic field is horizontal, or nearly so, detectable voltage gradients of the sort needed are not generated. Close to the sea surface these potentials range from 0.05 to 0.5 millivolts per centimeter, well above threshold for marine electroreceptors. In addition a shark should be able to sense electrically from such a gradient the speed and direction of a current in which it is drifting.

More important, sharks may have a magnetic compass that depends on electroreception of voltage gradients induced by their own rapid movement through the water. The potentials generated at its skin surface by moving through the magnetic field of the earth depends strongly on the compass direction in which the shark is swimming. Thus westward movement would generate positive potentials for receptors on its back relative to ones on its belly; eastward swimming would reverse this polarity. When the fish swims north or south there would be no voltage difference between their dorsal and ventral regions. This magnetic compass dependent on the horizontal component of the magnetic field would work best in tropical regions.

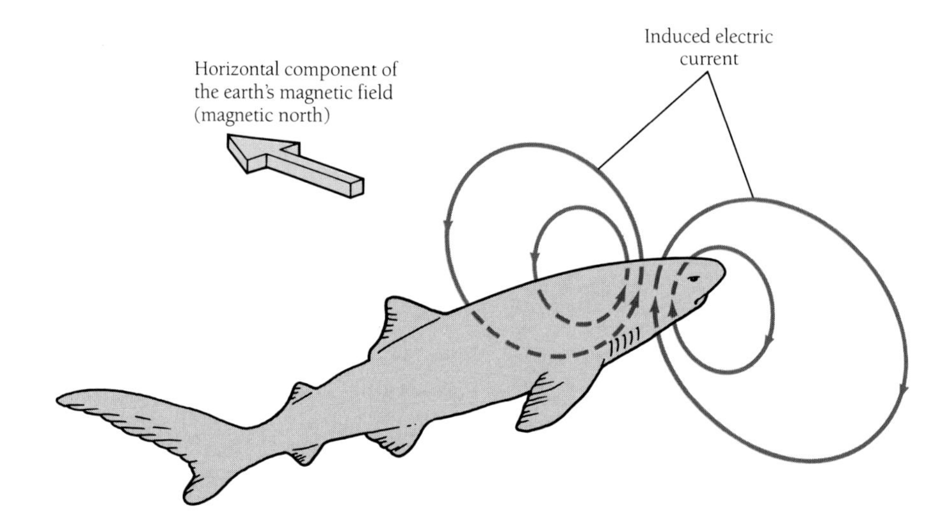

A possible magnetic compass in sharks depends on the stimulus pattern received by their electroreceptors as a result of swimming in the earth's magnetic field. Active movement to the east, as shown, would cause an electric current to flow upward through the animal and complete the circuit through the surrounding seawater. If the shark were swimming west instead of east, the currents would be reversed in their flow. In contrast, its swimming north or south would evoke no current by the mechanism suggested.

There is still another navigational possibility if the ampullary system can sense the electric current induced in a moving fish by the vertical component of the earth's magnetic field. The intensity of this current would depend simply on latitude and hence measure the animal's north-south location on the earth. Obviously these several proposed aids for shark navigation function through the particular ability of *marine* species to evaluate minute voltage differences between adjacent areas of their body surface. Freshwater electroreceptors are 10 to 100 times less sensitive. We know that sharks and rays can detect the stimuli postulated for these navigational devices. Whether they are able to navigate using such magnetically derived information about their direction and speed of movement remains to be proved by field experiments. Both the absence of known electroreceptors and the high resistance of the current return path through air argue against such electrically mediated magnetic perception by terrestrial or aerial animals.

Bacterial magnetic orientation

The first convincing evidence for a biological mechanism directly responding to the earth's magnetic field did *not* come from migratory animals. Instead it was found in microorganisms whose need for such an aid to navigation had never been previously suspected. In 1975 a biologist at the Woods Hole Oceanographic Institution observed magnetic responses in a common, but previously unidentified, marine bacterium that propels itself through the water by moving long hairlike processes called flagella. Within their single-celled bodies these bacteria have a series of magnetite particles closely lined up parallel to their cellular axis. These particles effectively constitute a single magnetic unit having north and south poles near the two ends of the organism. Such bacteria become aligned like a compass needle parallel to the earth's local magnetic field.

In the northern hemisphere where they were first studied, they move *northward* and *downward*, swimming along strongly dipping geomagnetic lines of force, which reveals that they have their south pole at their anterior end. Because they live in the mud, this oriented behavior provides individuals swept up from the bottom by currents with a perfect homing device for getting back down again. The orienting power of their internal magnet is so strong that even dead specimens are passively aligned. Even so, living bacteria are not passively dragged toward the magnetic pole but must swim actively to home toward the bottom. Magnetic bacteria in the southern hemisphere have their north pole at their anterior end. They move to the *south* and down to find the bottom. Near the equator, some bacteria swim northward and an equal number swim southward. The reversal of bacterial polarity from one end of the earth to the other makes this a magnetic device adapted to most of the global pattern of the earth's magnetic field. As proof that orientation of this kind has been around for a long time, geologists have recently found that several kinds of magnetite particles of bacterial origin are major components of deep-sea bottom layers dating back 55 million years.

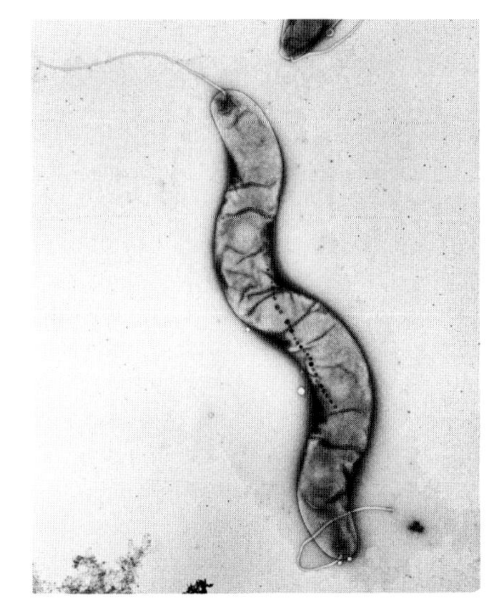

The marine bacterium *Aquaspirillum magnetotacticum* contains chains of magnetite particles, which appear as black spots in this electron micrograph. The chains act like the needle of a ship's compass, causing even dead bacteria to be aligned parallel to the earth's magnetic lines of force. The living bacteria swim by the long, hairlike flagellum, with their anterior pointed toward the earth's pole and downward.

Magnetite particles in other organisms

The compasslike behavior of magnetotactic bacteria with their known receptor mechanism made scientists wonder if magnetite particles might be found in other organisms. Perhaps the widespread reports of magnetic sensitivity in animals without apparent magnetoreceptors might be explained. A rash of discoveries in freshwater bacteria, in free-swimming single-celled algae (which have flagella for propulsion), and in honey bees, salmon, homing pigeons, and dolphins, among others, followed. Magnetite particles were found in many species, including mouse tumors, where they are quite unrelated to navigation.

For example, substantial numbers of magnetite particles were found in the skulls of chinook salmon, Pacific yellowfin tuna, and European eels. In these *large* animals the particles have to function quite differently from those in magnetotactic bacteria where they can physically orient the organism. Presumably in animals they must be linked to receptors such as the hair cells of the vertebrate vestibular and lateral line systems. Magnetite has been found in the honey bee, and its hair cell receptors apparently respond to weak magnetic stimuli as well as to gravitational stimuli. Magnetite particles have been found in the statolith of the guitar fish's sacculus. However, there is no evidence yet that natural magnetic forces acting on these particles could differentially stimulate the hair cells. In fact the required specificities and sensory tie-ins needed for these particles to be part of a magnetic compass have been widely sought but not found—except in microscopic algae, which quite resemble bacteria in this respect. Recent work on the bobolink promises some

The bobolink is one of the few American song birds that flies far across the equator on its long (10,000 kilometers, one way) annual migration. These birds nest in summer throughout southern Canada and the northern United States and fly east and south through Florida to wintering grounds in southern Brazil and northern Argentina. When the bird is exposed to light, the optic area and pineal organ of its brain become responsive to alterations in the magnetic field. Recent research demonstrates that the bobolink has still another system, which responds to changes in the magnetic field but is not linked to light. The latter may depend on a sensor containing magnetite particles and located internally between the eyes.

progress, however. Both fine magnetite particles and cranial neurons that respond to experimentally induced changes in the local magnetic field have been found in the birds' heads. Will physiologists soon discover a functional link between them?

Skeptical scientists who demand to see a bona fide receptor mechanism before acknowledging the existence of a biological magnetic compass have become impatient. On the other hand quite a few workers, as we have seen, are convinced that certain well-documented behaviors must be mediated by the earth's magnetic field. Although techniques of measurement and experimental control of magnetic stimuli have improved, much remains to be learned biologically. Well-designed experiments on animals in their normal environments, rather than in laboratories, are particularly needed to increase our understanding.

Other possible mechanisms

In addition to the magnetic granules already discussed, another suggested sensory mechanism involves certain biochemical reactions whose rate depends on the orientation of key molecules in the earth's field. Surprisingly this hypothesis involves vision and more specifically the critical molecule concerned is the visual pigment rhodopsin. The model requires that light absorption by a few of these molecules is controlled by the magnetic field so that the resulting stimulus pattern would provide compass information rather like the sun or like sky polarization. This proposed system, like the compass found in caged migratory birds and honey bee magnetic sensitivity, would provide axial sensitivity but not directional polarity.

Actually the first recordings of magnetic sensitivity in the central nervous systems of vertebrates showed that certain nerve cells and hormone-producing cells in homing pigeons, guinea pigs, and rats respond to weak magnetic fields in the presence of light. These responses to magnetic stimuli were obtained not from the retina but from the pineal organ, a dorsal appendage of the midbrain. In birds and mammals the pineal is not directly light sensitive but receives input from the retinas. Further evidence that the eye is somehow involved in the magnetic sense of the pigeon comes from recordings from brain centers for its vestibular organs. Some nerve cells there respond to light and also react specifically to the orientation of weak magnetic fields. To obtain such responses the eyes must be intact and the vestibular system stimulated just before testing. These conditions imply that some interaction between photic, gravitational, and magnetic inputs is required. Total darkness stops all magnetic sensitivity in these avian neural units, supporting the eyes' essential role. The optic centers and pineal organ of the bobolink also contain nerve cells that respond to external magnetic fields. Changes in the geomagnetic field also modulate the effects of directional light on dancing honey bee workers, suggesting that in insects, too, there may be some correlation between visual and magnetic sensing. However, the bee-dance errors affected by the earth's magnetic field occur in the normally dark hive. Clearly the question of multimodal inputs is fascinating and suggests the complex interplay of information sources needed for animal direction finding in nature.

To analyze animal direction finding, a major part of comparative sensory physiology and related orientation behavior has been briefly recounted in Chapters 5 to 7. Animals have evolved many intriguing direction finders based on all six of the sensory modes. Amazingly specialized mechanisms are common. The bat-detecting ears of certain moths, the sky-polarization analyzers in the dorsal rim of bee and ant compound eyes, and the receptors for female pheromones in the antennae of males of many moth species are cases in point. The most impressive means for global navigation are probably provided by celestial orientation and perhaps magnetic systems. Their more detailed recognition and understanding should have a high priority in future research. Particularly exciting are the recent electrophysiological recordings in birds, which may at last lead to a magnetic compass mechanism. However, as we already have seen in the Introduction, a compass by itself is of little value until it can be used to calibrate a map. So we must now turn to that topic.

8

Sense of Space: The Map

A sense of space depends basically on knowing the relations between the various places included in it. Hence the smallest bit of such awareness is the direction and distance from one particular place to another. A straight line connecting two points has both an orientation and a magnitude; mathematicians call such a line a *vector* and draw it as an arrow. The various compasses provide vector *directions* required for navigation. Thus a honey bee or an albatross heading out from its hive or nest can fly to a feeding place in the east, for instance, by using the known bearing of the sun to choose the right course. How such animals measure the *lengths* of the vectors is less well understood. Yet there is ample evidence that they can do so with impressive accuracy, perhaps by timing their travel at a constant speed. In any case a straight outward trip to a single feeding place would usually be followed by a flight home to store pollen or feed nestlings. Most simply this return would cover the same distance but have its direction inverted by 180°. Yet the actual homeward trip need not just follow the reverse bearing. Indeed with experienced adult animals, it may rarely do so.

FAMILIAR TERRITORY

Carpet of overwintering monarch butterflies. Every fall monarchs from the central and eastern United States navigate to the same few groves west of Mexico City. The precision of the feat implies that the insects have a maplike sense of space that is utterly astonishing.

In practice an animal's living space includes not just a single pair of places but rather many objectives and landmarks. The animal learns from experience the relations between home and various other locations, then uses this learning to find its way out and back again. The basic information consists of vectors radiating in all directions from a hub. As many as 75 to 100 vectors fanned out from the home atoll in the piloting repertoire of Pacific canoe navigators of the recent past. This sort of pattern is called a *polar* space sense in contrast to one having a *grid* of two reference coordinates like latitude and longitude. A mosaic of landmarks with irregular, but remembered, relations to one another could define another useful space sense.

Many foraging birds and insects use a polar space sense whose scale suits their needs and locomotor ability. Food collecting by a colony of honey bees provides an excellent example. Because biologists can read the dances performed within the hive, they can find out where the bees are foraging at any particular time. In this way the number and location of feeding sites can then be studied day after day. The insects' constantly shifting behavior shows that when food is plentiful, only the best sources are exploited. Poor patches are abandoned, scouts initiate exploration for better ones, and workers are recruited to forage on newly found rich areas. A given food source may not be persistently visited for more than a few days. Biologists learned that bees from a large colony in a woodland setting flew to feeding places, on an average, 2 kilometers from the hive; 95 percent of the sites visited were within a 6-kilometer radius around the hive. Hence the total foraging area was about 100 square kilometers. Within it workers visited about ten flower patches at any given time. Clearly the bees' collective perception of their surroundings can handle this food gathering plan in a highly sophisticated manner. The space sense involved is mainly a polar one with its hub at the hive and with courses to different feeding areas being vectors of various directions and lengths. Locations must be accurately stored as vectors in the insect's memory and an efficient sequence of visits must be paid to the current feeding patches.

Within the bee colony there is a clear division of labor with regard to navigation. Workers maintain a steady round-trip schedule to known high-quality sources.

Feeding places of a colony of honey bees were read from the bees' dances for consecutive days. The green dots indicate the feeding sites; the numbers are distances in kilometers from the hive at the center. Each day the colony used and evaluated many sites within 4 kilometers of the hive. Notice that the feeding locations changed, sometimes radically, from day to day.

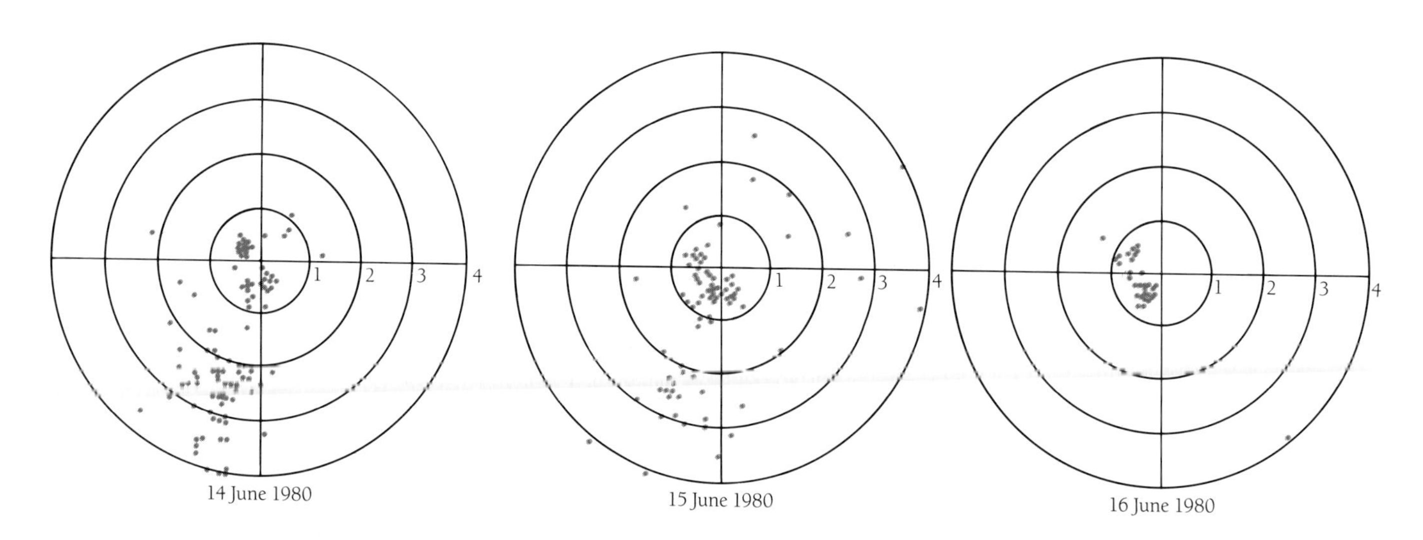

Some individuals specialize in nectar while others gather only pollen. Hence two different flight plans must be derived from the messages of dancing *scouts*. Additional distinct plans are needed if only one flower species is harvested by particular workers. The scouts explore systematically for new desirable feeding places. Depending on food abundance, from 5 to 35 percent of the workers act as scouts at any given time. They must not only find and evaluate promising areas but also on their return home induce hive mates to visit the new discoveries. Exploratory and searching behavior like this will be discussed further. First, however, we should consider how individuals forage.

Foraging details

Even for rather simple foraging, a single vector and its inverse or even a radiating array of vectors for various objectives will not account for detailed navigation. Thus when an insect or hummingbird has fed on one flower in a patch, it usually moves in a characteristic way to others nearby. Then, instead of returning home immediately it may visit additional feeding areas. To make the best use of its time and energy, the feeding animal must exploit the space concerned to obtain maximum benefit for a given effort. Bumblebees, for instance, which are solitary foragers, learn to visit sequences of feeding sites repeatedly. Individual honey bees are on their own, too, while they are busy collecting in a given flower patch. If more than one type of

Costa's hummingbird, a native of the American Southwest, samples nectar from these blooms. The foraging pattern of hummingbirds is similar to that of honey bees.

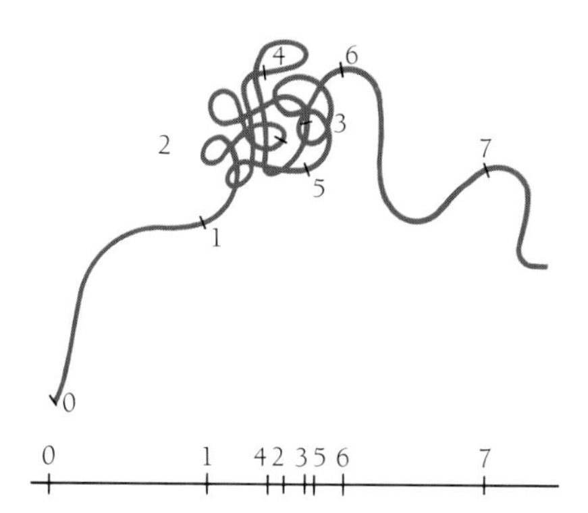

Some insects achieve an efficient foraging pattern by systematically changing the rate at which they turn. The path lengths between the numbers are equal, and the insect's speed of travel is constant. Yet the time spent in the food-rich area is markedly increased by greater turning rates (starting after 1) while the animal is feeding well. As it reaches the margin of the resource (at 6), its rate of turning decreases, and it moves out to find another worthwhile location.

blossom is present, particular workers may seek only the one whose scent or pollen was present during their recruitment dance and ignore the rest. Other flower-feeding insects and bats, as well as hummingbirds, honeycreepers, and sunbirds all show foraging behavior spatially organized rather like that of honey bees.

Local movements from flower to flower by insects of some species seem to be controlled through rate of turning in flight. When a given blossom is particularly rewarding, the insect's flight direction tends to change more rapidly. This is likely to keep it within the favorable area. If a flower just sampled has little to offer, the turning rate is reduced. This will carry the insect out of an unproductive region.

For predators, food searching is even more complex because prey are usually mobile and capable of hiding, fighting, or evasive action. The predator must take such behavior into account. Accomplished hunters like lions and tigers patrol trails followed by potential prey as do many other predators ranging from dragonflies to humans. Predators often encounter their quarry—not accidentally—at waterholes and feeding places. The brown bear learns to rendezvous seasonally with adult salmon migrating upstream to breed. Waiting at a riverbank or shallow stream, the bear skillfully snatches large fish out of the water, ordinarily with its mouth. Even though the animal obviously does not make use of spears or nets, its basic behavior is not unlike that of the traditional Indian fisherman in the American northwest! Among mammalian predators complex hunting behaviors of this sort are commonly taught to the cubs by the mother. She also helps to feed the young until they can hunt successfully enough to nourish themselves. Comparable behavior is well known in predatory birds like ospreys. Familiar territory and the animal's space sense are important here.

Storage and retrieval of food

Mammals that store food, like squirrels, the red fox, and the agouti, provide striking evidence for a vast spatial memory. They have the remarkable ability to recover hundreds or even thousands of pine nuts, acorns, or whatever else they may have hidden. Among birds, too, such space-specific behavior has been widely studied in the field and in laboratory experiments on various chickadees, jays, tits, crows, and nuthatches. Hiding and recovery of such items may take place locally within the bird's home territory or at sites several kilometers away to which special round-trips are necessary. Animals may hoard food briefly, between tides or for less than a day, for instance. On the other hand, the remembered location of stored food may be recalled for several months as when acorns buried in the fall are retrieved with striking precision during the following winter.

These various foraging and food recovery patterns prove that animals of many kinds have, or acquire and remember, extensive knowledge about their territory. Hence a neural record showing the relation between familiar persistent landmarks and new or temporary additions must be learned. Almost all carefully studied spe-

Grizzly bear catching a sockeye salmon in the Brooks River, Alaska.

cies show marked changes in their space sense from one stage of development, to another, such as from larvae to adults, and also with ongoing experience in their environments. In previous chapters we have seen that juvenile and naive individuals may orient and behave differently from seasoned adults. An important feature of a *mature* animal's familiarity with its territory is that it appears less dependent on a polar system made up only of vectors radiating from a hub. Even when travels begin and end at a central spot the ability to visit a network of other locations by various pathways is clear evidence of a more complex spatial memory. We have seen that precise returns of migratory birds require previous familiarity with territorial details. This implies that long-range navigation locates the familiar area while the learned neural map provides detailed guidance within it.

PIGEON HOMING

The ultimate test of a space sense is whether it will enable an animal to move toward a known goal from a place that it has never visited before. This is a feat that the homing pigeon achieves regularly. Claims that certain species of wild birds including the ring-billed gull and the wood thrush may possess this ability have been made, but the data are few. The homing of the domesticated pigeon after being displaced has provided one of the most popular models in the recent study of animal navigation. Domesticated pigeons and wild rock doves belong to the same species *Columba livia*. The wild birds are nonmigratory and not known to have any unusual skills related to orientation or direction finding. Yet since ancient Egyptian times, domesticated "carrier" pigeons have been used to bear messages home from wherever the birds had been transported. Particularly useful in old-fashioned military campaigns, this practice went on for several thousand years until replaced by the telegraph in the mid-nineteenth century. Since then the training and selective breeding have been widely continued, particularly to sustain the sport of pigeon racing, which produces some extraordinary feats. Several-thousand-kilometer homing ranges have been authenticated, with maximum speeds up to 145 kilometers per hour and 24 hour runs of more than 1300 kilometers. In the annual Belgian contest the birds race from Toulouse in southern France to Brussels, a distance of 750 kilometers.

Against this practical and sporting background, biologists have been trying for some decades to determine the mechanism by which trained pigeons are capable of such remarkable feats of navigation. Particularly active centers for this study have been Cornell, Duke, Frankfurt, Pisa, and Tübingen Universities. We have already considered a number of relevant sensory systems that seem to be involved as compasses. The main new point to be emphasized here is that adult pigeons can home *several hundred kilometers* from locations where they have never been before. These mature, experienced birds must have the extraordinary skill of determining the direction of home from the site to which they have been displaced. Homing success varies with specific localities and with season. It even has been inversely correlated with sunspot activity, which creates magnetic storms.

Such homing would be impossible by piloting unless the animal could somehow identify landmarks or seamarks without previous direct experience. Human pilots do this with the help of published charts showing geographical details. Animals could presumably do so, too, if their space sense were genetically determined or, if they are social, it was communicated to them by parents or other more experienced individuals. Aside from piloting, the animal might be using dead reckoning, celestial navigation, or geophysical techniques of some sort. Juvenile and naive birds can use the earth's magnetic field and the sun as compasses but a well-integrated place sense develops only with extensive experience or training.

A puzzling aspect of pigeon homing remains to be mentioned. Upon release from previously unvisited places, pigeons often start flying in a direction quite different from that toward home. Usually, in experiments concerned with homing, an important datum is the course on which a pigeon, under observation with binoculars, disappears from sight. The initial compass headings of all the pigeons in an experiment, plotted and analyzed in a circular distribution, frequently show that the average bearing near the release point is quite different from the true home direction. Substantial errors of 30 to 60° clockwise or counterclockwise from the direc-

The electrical coil strapped to this pigeon's head alters the intensity of the earth's magnetic field in the area around the bird. Homing experiments using these coils demonstrate that domesticated pigeons can use the earth's magnetism as a compass, but only trained birds have a well-developed sense of where they are on the "map."

tion to the home loft are not uncommon. They typically depend on the specific release site and presumably on some of its particular navigationally significant features. Thus the error a bird makes on being released 50 kilometers *east* of the loft typically differs in both size and direction from that it makes when released 50 kilometers *west* of it. Such deviations also differ for lofts in different locations. Some local, presumably geophysical, clues must guide the pigeons' course as they fly away from the release point. Careful study has shown that whatever such geophysical effects may be, their influence varies with time of day and level of sunspot activity. The resulting detours may contain hints, as yet undeciphered, about the pigeon's map sense.

Curiously, homing success and accuracy seem not at all influenced by these initial "mistakes" in heading. Following the birds in a light plane, biologists have found that a gradual correction smoothly reduces the original course error toward zero as the birds fly home. Although such navigational behavior cannot be adequately explained, some speculation is possible. For instance, to get on the true homing course the animal may need to integrate or evaluate its map and compass information over more time than it takes to disappear from the observer's view. Alternatively the initial geophysical input to the sensory compass may gradually be replaced by better information that becomes increasingly available as the bird proceeds. In any case pigeons in homing seem to depend mainly on a compass heading rather than visual piloting, except very close to the loft.

PILOTING AND THE MAP SENSE

A course navigated by piloting can be one vector or a sequence of vectors leading from one recognized geographic point to another. Any image-forming sensory system could enable an animal to perceive landmarks or seamarks for piloting. Animals express their sense of space most obviously in visual piloting. As has been mentioned, many migratory birds have geographically distinct flyways along which they travel seasonally. Not only will a given species follow its own route year after year, but the same corridors may be used by a variety of species as different as ducks, finches, and hawks. That such migration paths are often associated with conspicuous landmarks like a shoreline, river valley, mountain ridge, or pass is certainly suggestive. Whether these features function as route maps or as barriers is not certain, however. The commonly observed funneling effects of geography, like the influence of the Iberian and Florida peninsulas on southward migrators, are real enough. But the question of whether the animals are following the shorelines as guides or just avoiding flying over the sea until the last possible moment is not so easily answered. Some data imply that birds may use an extensive landmark to compensate for wind drift yet not be guided by it when the air is still. Migratory routes cannot depend solely on the availability of landmarks. Many species must

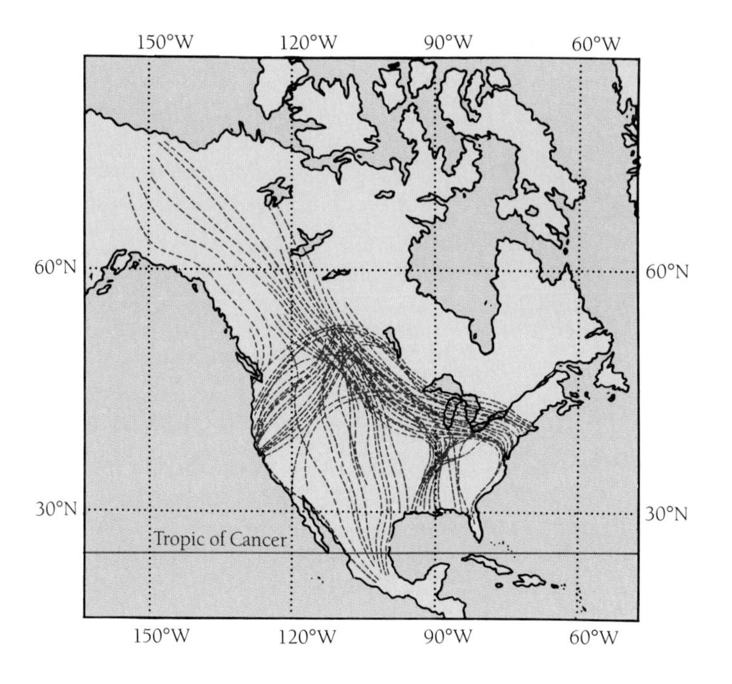

Principal flight paths followed over North America by migrating canvasback ducks. Narrow routes, such as the Mississippi Valley, exist along with broad paths from northwestern Canada to Mexico and across the northern continental United States and southern Canada.

stop to feed and rest along the way. Routes that have evolved bring birds to suitable stopping places. Furthermore many species with widely separated breeding and overwintering areas move between them on a broad front rather than along a narrow route. This seems particularly so for nocturnal bird migrants. For such migrators to use piloting, there would have to be many local sets of landmarks, all visible at night. Also, a considerable number of navigators, including small land birds, regularly undertake long-distance journeys over water where there are no visual geographical clues, except for wave patterns, for hundreds of kilometers. In such cases piloting by eye must be replaced by other navigation techniques like those afforded by the earth's magnetic field, the sun compass, star patterns, or sky polarization.

Some insects certainly navigate by visual piloting based on detailed familiarity with landmarks within their territories. The behavior of honey bees strongly suggests that they first systematically learn the features of local areas and later use them in navigation. To test which features these insects can learn to recognize during their exploratory flights, researchers have introduced artificial landmarks. When these are repositioned or removed, the worker bees stop foraging and make new orientation flights before they resume feeding. Similarly, when they visit a new foraging area for the first time, they make comparable reconnaissance flights within the site as if they are memorizing the terrain before homing. An important interaction between such remembered landmarks and the celestial path of the sun has been demonstrated in two species of honey bees. On fully overcast days workers show by their dances that they have memorized the sun's positions in the sky relative to known local landmarks like trees or hedges. Furthermore, in using that information, they compensate, at least roughly, for the sun's movement with time.

Among nonvisual senses used for piloting, we have already discussed the acoustic "image" of echolocation, thermal mapping by infrared receptors, the close touch sense of the lateral-line system, the olfactory clues used by homing salmon, and the chemical orientation proposed for homing pigeons. Still others are electric field perception, which permits electrolocation, as well as the use of specific magnetic and gravitational landmarks. Remembered geographical relations are available in any of these sensory modes.

Whereas earlier researchers often presumed that there exists a single general solution to problems of animal navigation, the current view takes for granted that more likely there is no simple *one*. Rather, different techniques, sometimes in a preferred sequence, are used by the same individual under particular circumstances or during various stages of development. A given species may have its own distinctive ways of navigating different from those of other species. Even individuals of the same species living in different places and having different experiences may behave in their own characteristic ways. As with direction finding, geographical relations are almost certainly perceived in practice by using all possibly helpful information.

Migrating animals must have a space sense that is greater both in scope and content than those of nonmigrating animals. Typically the space sense of long-range travelers includes at least three components—two subhabitats plus a pathway connecting them. For instance, arctic terns have a familiar territory in the summer breeding area. Their main winter territory for adults near the antarctic continent and south of the Indian Ocean is shaped by strong prevailing westerly winds and by the localized abundance at the ocean surface of organisms on which they feed. Within this area they are nomadic, apparently having no fixed roosting or feeding sites. Global pathways connect their high-latitude northern and southern territories. When single birds traverse these pathways for the first time, the series of vectors followed must be based on genetically built-in reflexes. First-year juvenile terns do not depend on the leadership of adults and thus seem to follow an innate—rather than a learned—space sense.

Whether learned or innate, an animal's space sense may be thought of as its *map sense* by analogy with human aids for place finding. For us a map is graphic shorthand that represents spatial relations. It may be a fairly realistic representation of land and sea. More usually it is quite stylized with numerous symbols or conventions used to represent real relations such as land elevations or ocean depths. Details of this sort usually vary with scale, the specific terrain involved, and the particular use for which the map was made. Thus a geological map for mining prospectors differs considerably from one of the same region showing airways and flight patterns for plane pilots. Most maps represent the nearly spherical surface of the earth as a plane. The larger the area, the greater is the distortion resulting from this conven-

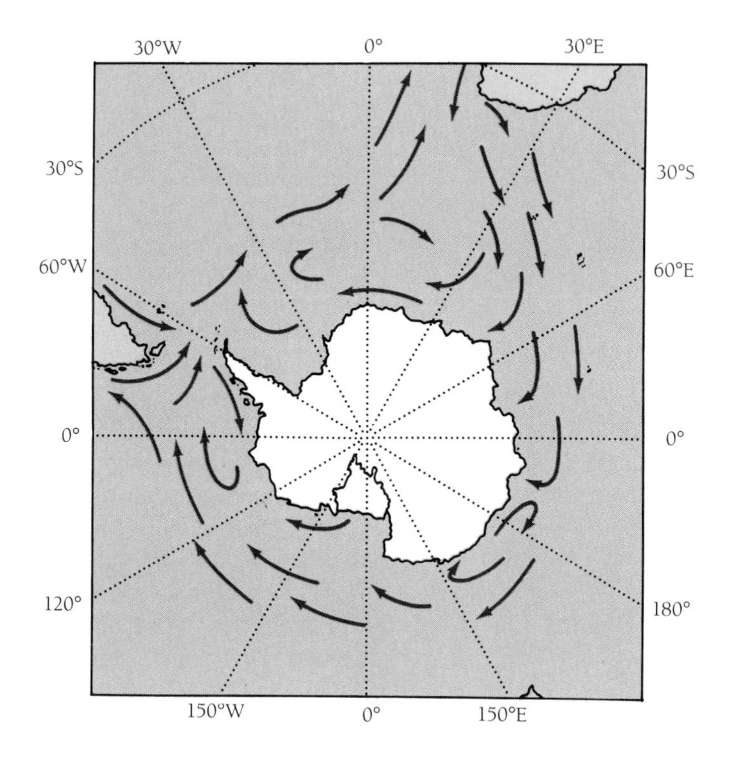

A polar map of the Antarctic. The lines of longitude radiate out from the pole and intersect circles of latitude. Arrows show the migration of the arctic tern toward this southern hub, around it, and back out.

tion. Many maps have two coordinates, usually latitude and longitude, that appear nearly rectangular at moderate latitudes. Recall, however, that close to the poles these conventional mapping coordinates become straight radial lines of longitude and circles of latitude; the grid map becomes a polar map. Although navigation is possible in any direction through 360° one can only travel north from the South Pole! How does the space sense animals clearly possess relate to our own charts and globes? Obviously no animal has written diagrams or physical models but the honey bee's dance patterns, which specify directions and distances, seem close to being a polar map. Other social animals may also communicate maplike information.

THE LAYOUT OF ANIMAL MAPS

Whatever its modality, the basic geometry of animal maps is a matter of great interest but little certainty. A major dilemma is whether the space sense is one or two dimensional. Is it essentially *linear*—like a vector or a sequence of vectors—going

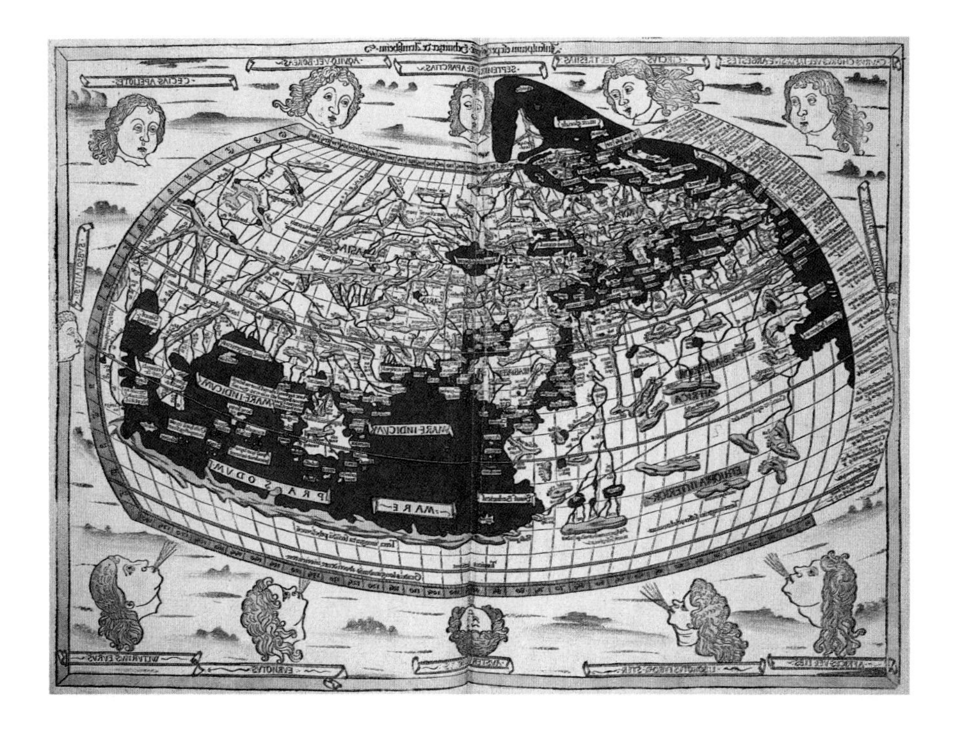

A Ptolemaic map produced in Ulm, Germany, in 1486. Ptolemy, a Greek geographer of the second century, used a two-dimensional grip of latitude and longitude to locate any point on earth. When Ptolemy's atlas was rediscovered during the Renaissance, its grid system became conventional for western European maps.

from one starting point to a given destination? This is like a human route map indicating only how to go specifically between *A* and *B*. Once off the route, the navigator with such a map is lost. Or does the animal's space sense provide a two-dimensional image? The polar space sense, for instance, combines a number of individual route maps radiating out and back from a hub such as a bee hive. A two-dimensional map could also be a *mosaic* in which each component is recognized as a landmark with fixed relations to all the others. Finally a two-dimensional map could be a *grid* map with two intersecting gradients like the latitude and longitude standard for human maps. Mosaic and grid maps allow a wide choice of routes within their areas and a grid map could permit an animal to fix its position. A linear or polar map is relatively simple but inflexible; a mosaic map and a grid map are more versatile and more complex. No doubt animals have both linear and two-dimensional types, but evidence suggests that route maps are more common. Experimenters have shown that the space sense of desert ants is of the route type. Even a few meters of imposed relocation, well within the animal's normal foraging range, results in a corresponding error in its attempt to reach home. Similarly, displacement of inexperienced migrating birds, like juvenile starlings, evokes a corresponding mistake in their overwintering location. After a significant shift in location, resuming travel on a given compass heading without adjustment must lead to navigational errors.

The behavior of bees suggests that at certain times they depend on route maps. For example, bee keepers know that after a hive is moved only a few meters, a confused cluster of veteran workers will buzz around the space where the hive entrance had been! Such localization apparently depends on piloting by landmarks remembered as a series of "snapshots." Landmarks have been shown to function as direction finders in the dances of the Asian bee *Apis florea* much as celestial clues do for the European honey bee. Bees may have more than one type of space sense, however. Evidence for a map more generalized than a route map has been recently reported in the honey bee. Still another example of animals' behaving as if they are following a route map was seen in migrating salmon lured to swim up a stream artificially labeled with a chemical to which they had been conditioned. Clearly a simple, but specific, olfactory route map was being used to guide the animals' directional choices.

Although the mental maps of Pacific canoe navigators as well as those of a variety of animals are polar, most charts used by sophisticated human navigators provide numerous landmarks and seamarks; they also have rectangular coordinates, which allow any conceivable voyage or flight within their scope to be planned conveniently. The pilot using the map need never have visited its area at all. In fact, trips to certain combinations or sequences of ports of call may never have been previously made by anyone. In addition, a gridlike space sense should allow an animal, which has been accidentally or experimentally displaced within its territory, to recognize sites where it has never been before. Also such a two-coordinate map might permit an animal to reach a given goal by a path not previously tried.

A sextant is being used to locate the north pole. Looking through the instrument, the navigator measures the angle between a celestial body and the horizon.

Position finding

For an animal to return from a previously unvisited location to a known one, like home, two quite different methods would seem to be available. In one, rather abstract knowledge about locally perceivable clues permits places to be recognized directly. For such a place sense, some kind of orderly coordinate system or known mosaic of features would be needed. Using regular coordinates or gradients, an animal could identify a new location and relate it to its goal. Humans, for example, can, with instruments and tables of data, precisely and uniquely define any particular place on earth by its latitude and longitude. A sextant and chronometer can be used almost anywhere under clear skies to fix one's unknown position with an accuracy of 2 to 3 kilometers. *Place-specific* celestial information is used in this way to determine location.

Starting with a defined position or *fix,* it is routine, for the pilot or captain using global geometry, to plot the shortest course from that point to any other place on earth. The plotted course, linking the fix to the goal, consists of one or more vectors whose directions and lengths are described by the coordinates. Is it possible that animal navigators could conceive and make use of such a system? Except for the pigeon and other birds already mentioned there seems to be little solid evidence that any other animals have such an abstract site-specific ability.

The other way to identify present position, said to be *route specific,* is to use data acquired while moving from a known location toward a previously unvisited position. For this, dead reckoning, including inertial navigation, as well as any other specific information such as sights, smells, or sounds perceived during displacement or on the outward journey are used to make a fix on the new site. Because so many alternative types of data are possible, it is extremely difficult to conduct experiments proving that an animal does or does not reach a goal by following remembered local signs. Experimenters must eliminate all potential clues as they move the animals to new places by reducing the animals' sensory input as close to zero as possible. Accordingly, animals have been blindfolded, kept in total darkness, shielded from magnetism, exposed to distorted magnetic fields, and rotated on turntables to prevent displacement-derived information from contributing to a fix. Researchers in Frankfurt have shown that young inexperienced pigeons may lose their ability to home when subjected to such treatment on the outward journey. Yet adult, trained birds, treated in such ways home as well as control birds transported without the sensory deprivations.

Either a coordinate grid or a memorized mosaic is needed to define a point on a surface. One of the usual human map coordinates, *latitude,* might be locally perceived by animals from the dip of the earth's magnetic field. Latitude also might be perceived from the sun's elevation at noon, the elevation of the polestar, or recognition of stars that reach the zenith above particular places. But determining *longitude* requires timekeeping of such accuracy that it was not possible for humans at sea until the late eighteenth century when Harrison's chronometer, an instrument that could keep extremely precise time over long periods even on shipboard, was built. The latter provided an accurate time standard to compare with celestially observed local time. Because of the earth's fixed rate of rotation, each hour of such time difference equals 15° of longitude from the reference location as reflected in world time zones.

Can animals possibly have such a reference? If not, how can they precisely sense a second global dimension, ideally forming a perpendicular grid with latitude? Even if an animal navigator could fix its current position, how could it know which way and how far to go to home, or to another objective? A neural map large enough to include both locations in the animal's memory would seem to be needed. If such a map were available, just a compass would suffice to provide the direction required from the fix.

Animal map coordinates?

Biologists have proposed several possible coordinates that might serve for animal fixes. In one hypothetical system, devised in 1947, two geophysical forces account for a bird's sense of place. First, homing pigeons are assumed to sense latitude from the Coriolis force, which is due to the earth's rotation. This might be possible because the force increases as the distance of an animal from the earth's spin axis decreases with the bird's movement north or south of the equator. The hypothesis assumes that pigeons can sense the differences in force intensity and thereby fix their north-south location. Second, the birds are postulated to sense longitude from the deviations of the earth's magnetic field from north-south directions. This would require that the pigeons compare magnetic force orientation with a geographic direction. The intersection of these two coordinates could provide the birds with a fix, whether they had ever been in that place before or not.

Another proposed mechanism depends on celestial data. This is the sun-arc hypothesis of Matthews, which requires highly accurate perception of the solar path through the sky combined with precise internal time measurements to yield the animal's locality. However, on experimental as well as theoretical grounds, both of these hypotheses have for some time been judged quite unlikely. Some other rather intriguing possibilities, on the other hand, have not been properly tested or perhaps even deemed worthy of it. For instance in some relevant parts of the world, like western and central Europe, the lines connecting points of constant magnetic dip and the meridians of equal magnetic deviations from true North form a nearly rectangular grid geographically. But to date there is little evidence on whether animals can use or even sense these features. A different pair of magnetic coordinates might also form a global map: for instance dip and field strength—if they are perceivable and mean something to the organism.

But do navigating animals have a two-coordinate map sense?

Spatial memory and neural maps

Learned geographic data must be directly perceived by each individual animal and somehow retained for longer or shorter periods in its central nervous system. Psychologists define such remembered spatial relations as "cognitive maps." Animals may gain access to such neural records as needed for behavioral use. Where and how such information is kept are obviously challenging questions when asked regarding human behavior. They are even more so when we apply them to the many types of animal navigators. Clearly they cannot tell us directly about their geographic memory, and indeed we cannot prudently surmise beyond their observable behavior to what extent they are even aware of such vital data. Hence researchers need a strong combination of keen insight, experimental skill, persistence, and good luck to make progress on this topic. The difficulties are multiplied by the variety of kinds of maplike data that may permit different modes of navigation whether piloting, dead reckoning, or using celestial bodies. Furthermore, all the senses used for direction finding, as well as their main subdivisions like hearing and the lateral line, could each have its own spatially organized sensory terminals in the brain. Altogether then, there may be a rather astonishing array of animal cognitive maps!

In general, the nature of such maps can only be inferred by studying natural behavior and analyzing it as best we can through the various sensory inputs and information-processing systems involved. Physiologists have determined—at least for the vision of arthropods (insects, crustaceans, and their relatives) and for several senses such as hearing and vision, of vertebrate animals—that maplike projections of various features of the external world are present in the nervous system. Intuitively this systematic organization of spatial data would seem to make sense. Similarly, connections between different sensory channels providing parallel data on the same part of the environment are perhaps not surprising. For instance, we saw that in the brain of certain snakes the projections from sensing infrared and visual space fit together intimately as do those from electrolocation and vision in certain fish. In such cases the central network for information processing must operate to find similarities and inconsistencies in the multiple inputs. Clearly, orderly mapping of nerve-cell terminals should simplify this so that the most reliable and relevant information governs the animals's behavior.

There is good evidence that the hippocampus, part of the posterior forebrain, is a key element in enabling mammals to learn and remember the geography of their living space. A rat, for instance, probably improves its maze running as the selective transmission of stimuli among the cells of its hippocampus increases. This brain center apparently interacts with the cortex in maintaining the cognitive map. The underlying mechanism of learning in general and the functional basis of memory have been topics of major biological interest for several decades. As a result, the rather extraordinary capacity of certain animals, like cockroaches, birds, and rodents, to acquire detailed central nervous system maps of their habitat has attracted considerable attention. This has come not only from comparative physiologists but also from ethologists, ecologists, and experimental psychologists.

Many studies of vertebrates have shown that, as these animals learn, the microscopic structure of their central neurons changes in certain ways. Although, as we have seen, some invertebrate animals certainly do have extraordinarily good spatial memories, few similar studies have been made on these species. Biologists have found, however, the neuron structure in the so-called mushroom bodies of the bee brain undergoes marked changes after the insect's initial exploratory experience in flying. Experimenters observed this by comparing these neural centers in preflight juvenile honey bees about a week old with those in similar individuals that had made only a single orientation flight away from the hive and back. Even after such an apparently simple learning experience, specific neurons showed persistent swelling of certain parts accompanied by shortening of other parts. The changes observed might be expected to lower resistance to information transmission between the cells involved, as they have been proved to do in certain cases for mammals. Mushroom

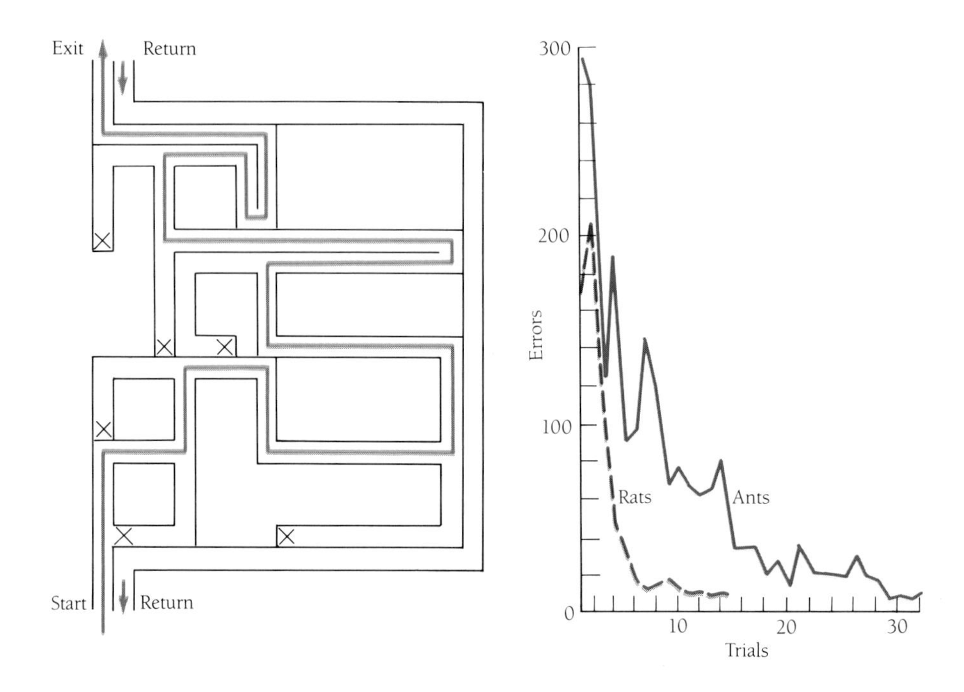

Rats and ants can learn to run directly through mazes, avoiding blind alleys. In the maze drawn here, the animals had to learn six turns to avoid blind alleys. As the graph shows, the rats reduced their errors in five or six trials and made few wrong turns at all after they had tried ten or twelve times. Ants, too, became able to run the maze, but they learned less quickly and were still making some errors after 32 trials.

bodies in some insects have been associated with memory, particularly for olfaction, and are known to receive sensory input from several receptor types. Hence it is tempting to think that the neuronal changes documented after the first territorial learning by the honey bee might indeed "be" the map that is developing! Cell biologists, however, are beginning to find other signs of both short-term and long-term memory in cells and even in particular molecules within cells. Membrane receptors and chemical transmitters as well as the mobilization and synthesis of enzymes and messenger RNA underlie the growth and other changes correlated with learning in nerve cells. Obviously, information on the geographical relations between places is basic to any navigation. Yet that it coexists somehow in the memory of animals as well as in their genes and oriented reflexes makes their map sense notably difficult to master. The effort should be all the more challenging for that!

SIMPLE SOLUTIONS

Perhaps animal navigators—rather than having skills roughly analogous to trigonometry, cartography, and map reading—depend on certain built-in mechanisms that demand relatively little sophistication. For instance, procedures that are simpler and less anthropomorphic than a two-coordinate map sense may enable them to home from a previously unvisited location. A few simple "rules of thumb" that operate as reflexes might provide adequate guidance even for apparently complex oriented movements. In such a case some sort of reflex mechanism could play a major role. Recall the rather simple navigation program proposed for the extended fall migration of many passerine birds, wind driven over the western North Atlantic Ocean to Caribbean islands and South America. Another excellent example of such a rather simple built-in device for navigation can be seen in the way the honey bee and the desert ant steer by sky polarization. Recent evidence suggests that both of these insects have evolved a rather elaborate filter for this purpose in a limited area of the compound eye. The functional units in this area are so arranged that only when the animal's longitudinal axis is nearly horizontal and parallel to the sun's bearing, will the polarization of the clear sky stimulate both eyes maximally and thereby provide a compass direction to steer by.

As we have seen, this compass depends on ommatidia along the dorsal margin of each eye, having axes of polarization sensitivity oriented like a fan. Using this filter the insect can routinely establish its bearing from the sky's polarized light. Compared with other proposed explanations, this method supplies a rather *simple* steering reference. Sometimes its use when only restricted areas of blue sky are visible results in substantial directional errors, but generally the system is accurate enough to be practical. Remember though, that to find the food source the bee still needs to learn from a scout the angle between the solar bearing and its goal. How all this information is used to derive the actual direction to be steered is still largely unexplained.

Another simple navigational mechanism has been proposed to explain how magnetic sensitivity could relate to the global migratory routes of birds or other animals that can sense the earth's magnetic field. The model requires simply that the animal assess the dip of the earth's field as it starts to migrate in a particular geographic direction and thereafter simply maintain a course that holds the magnetic dip constant at its initial value. A number of well-known migratory routes for birds seem to fit this mechanism well but there are other cases that do not. Consequently further study of this hypothesis and particularly some experimental evidence for its practical significance are both clearly needed here. At the moment a simple reflex of this kind might seem to explain too much!

EXPLORATION AND SEARCHING

An interesting point about spatial memory is the possibility that *exploration* may often be an effective means to widen and reinforce an animal's map. It has even been argued that exploration followed by travel over remembered terrain are the main components of animal navigation. Long-range movements and migrations in this

These northern shoveler ducklings are starting to learn the map of their territory by exploring their immediate neighborhood.

view are just greatly extended trips within an individual's known territory. This appears to contradict the well-known fact that many first time migrators make independent global movements far beyond any earlier experience. It also ignores the extensive evidence for navigation by means other than piloting, such as using celestial clues. In any case exploratory behavior, like juvenile play, could indeed serve an animal for collecting data or offering experiences that might be useful later. Honey bee scouts, for instance, not only explore for patches of flowers but also scout for optimal sites for a new hive. Exploration implies invading fresh areas previously unknown to discover new resources. Hence it involves quite generalized objectives and activity patterns. Note, however, that successful returns home after forays into new territory are essential for survival unless the animal is completely nomadic. As a result the outward journey must be remembered in enough detail to make homing reliable. This means that the animal's internal calculated location relative to its starting point must be constantly revised at each step in its exploration, just as in human dead reckoning.

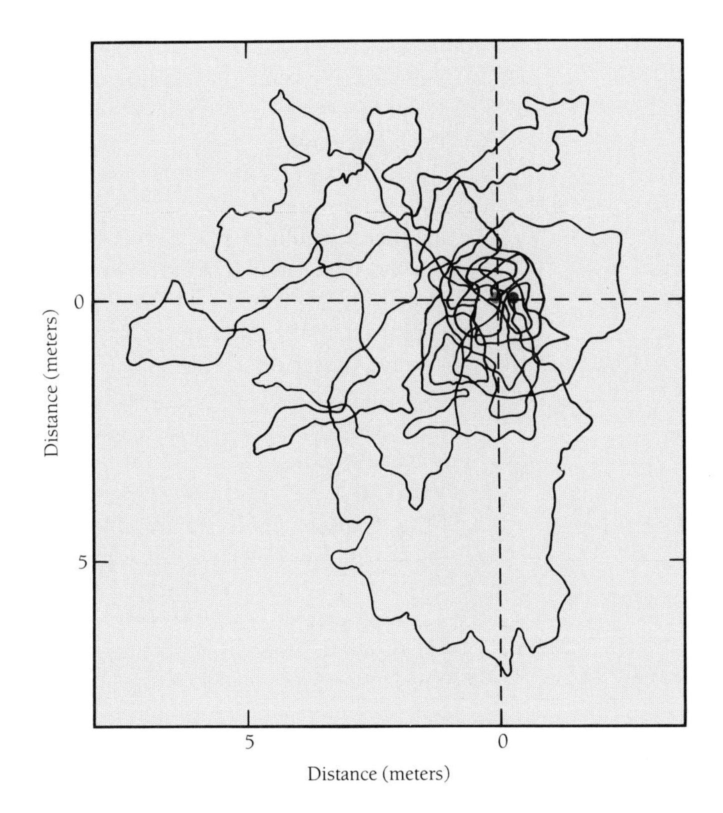

A foraging desert ant meanders for 21 minutes over an area of about 120 square meters from its starting point (green dot) to its ending point (red dot). If disturbed on course, the ant can home directly and accurately. This ability must depend on some kind of ongoing dead reckoning or vector addition along its way.

We saw that the desert ant *Cataglyphis* forages by taking a meandering exploratory path away from the nest. Nevertheless the ant at all times knows from dead reckoning just where its unseen nest hole is and can return to it in a straight line using sky polarization. Columbus on his voyages had a similar problem to solve in reaching his home port after bold, remote exploration. This he repeatedly solved on his several major voyages using the rather crude ship's navigation then available. *Searching* in contrast to exploring includes hunting for something familiar that has become lost or is evasive within a familiar area. So Columbus, while hopefully searching for Japan and east Asia, was in fact exploring the West Indies whose very existence was previously unknown. In any case both exploration and searching may well use quite similar techniques for covering territory systematically and efficiently, as well as for returning home.

What happens when an animal's navigation is inadequate or fails? How does it cope with mistakes or accidents, which must occur fairly often? Obviously, unpredictable winds and currents may cause an animal to miss its goal. Severe storms may have disastrous effects on bird and insect migration as we saw they did for canoe navigators. Floods, droughts, dams, and pollution clearly affect aquatic forms. Do animals have some general modes of behavior for getting home or back on course when their map or compass proves inadequate? In some cases at least, the answer to this question is certainly "yes," and the recovery behavior may be surprisingly effective. Early reports that the flights of displaced birds can resemble organized searches were made in the 1940s by a zoologist who followed the paths of displaced gannets and herring gulls in a light aircraft. He observed that birds transported and released beyond their familiar territory flew successively in various directions in apparently systematic attempts to find their nesting sites. Obviously regular searching, for instance on radial or spiral paths, would lead to familiar territory if the searching covered ranges on a scale similar to the displacement. In a sense a search plan is a navigational recipe to use when an objective has been lost and more direct procedures have failed. Evidence for such remarkable adaptive behavior has more recently been clearly demonstrated in arthropods such as the desert ant. Eels, too, have been found to home reasonably well, either by direct navigation or by searching, after being displaced as far as 100 kilometers in the North Sea. However, nest homing known in some displaced passerine birds seems, even when the return is slow, to be direct rather than via a search pattern.

The foraging behavior of the desert isopod *Hemilepistus reaumurii* is rather similar to that described for the ant. This terrestrial crustacean also has been found to navigate as if by dead reckoning and celestial direction finding in order to return home. However, its orientation is not accurate enough to localize the small nest entrance directly. Usually the isopods complete their near return by searching in an apparently complex and systematic way after arriving close to home. This search behavior starts typically with a widening spiral course that is followed by a series of radial loops in various directions. A random search seems to be combined with a tendency to favor the sector deemed most likely to contain the goal. Desert ant foragers, displaced when they have almost reached their nest hole, have also demonstrated remarkable search patterns. These consist of a series of exploratory loops centered about the starting point to which the ants repeatedly return. The length of these radial forays increases systematically, following some probability function. Their direction varies randomly. Amazingly, a model developed to account for this ant behavior closely resembles an ideal search pattern previously devised by United States Navy scientists to aid hunting for a lost object such as a disabled submarine! The search method of both isopod and ant seems well adapted to locate home quickly. As the desert ant forages, its meandering traces out a rather similar pattern, but it makes the largest loops first, decreasing their length as it continues. Many other arthropods including spiders apparently have similar systematic patterns for searching.

On homebound trips, the terrestrial crustacean *Hemilepistus reaumurii* navigates as if by dead reckoning and celestial direction finding to the neighborhood of its nest. It then uses searching behavior to find the nest hole.

However, "random walks," defined as paths made up of a sequence of vectors varying randomly in direction and length, have been evoked repeatedly to explain effective homing or migratory routes in a variety of animals. For Pacific salmon, such a completely unstructured searching was found to be inadequate to account for the known return of fish tagged at sea. In other words, some correct sense of the home direction is needed in the fish's swimming behavior. However, according to a computer model, adding even a slight directional bias favoring the right homeward sector to a random walk might account for the observed level of homing success. This necessary bias may, of course, be viewed as a weak version of the minimal unit of map information—a single vector directed toward the goal—with which we started this chapter.

9

Sense of Time: Biological Clocks and Calendars

Rhythms pervade living things. Nerve impulses, heartbeats, locomotor movements, breathing, sleep-wake alternations, reproductive cycles, the span of generations, and oscillations in population size—all witness the familiar periodic nature of life. The timing ranges from milliseconds to years or decades. Our working definition of migration specifies it as a periodic movement from one part of an animal's range to another and back. Its timing is a characteristic feature. Navigators of all kinds measure and use time. Many animals, as we have seen, have a sun compass that corrects itself for daily movement of the sun through the sky. Quartz crystal timers, atomic clocks, and satellite time signals nowadays make precise time-keeping routine for sophisticated human navigators. How do animal navigators keep time? We will learn in this chapter that certain periodic events in the organism serve as an internal master clock that is constantly reset to fit external natural cycles like the solar day.

INTERNAL RHYTHMS

Survival and successful reproduction usually require the activities of animals to be coordinated with predictable events around them. Consequently timing and rhythms in general are vital when biological functions must closely match periodic events like time of day, the tides, lunar cycles, and the seasons. The relations between animal activity and these periods have been of such interest and importance, particularly for daily rhythms, that a huge amount of work has been done on them, and a special research field of *chronobiology* has emerged. Normally the constantly

The California beach is covered with these silvery, squirming grunions, which wash up on the sand to spawn. Spawning is timed to coincide with a nighttime high tide 3 or 4 days after a new or full moon.

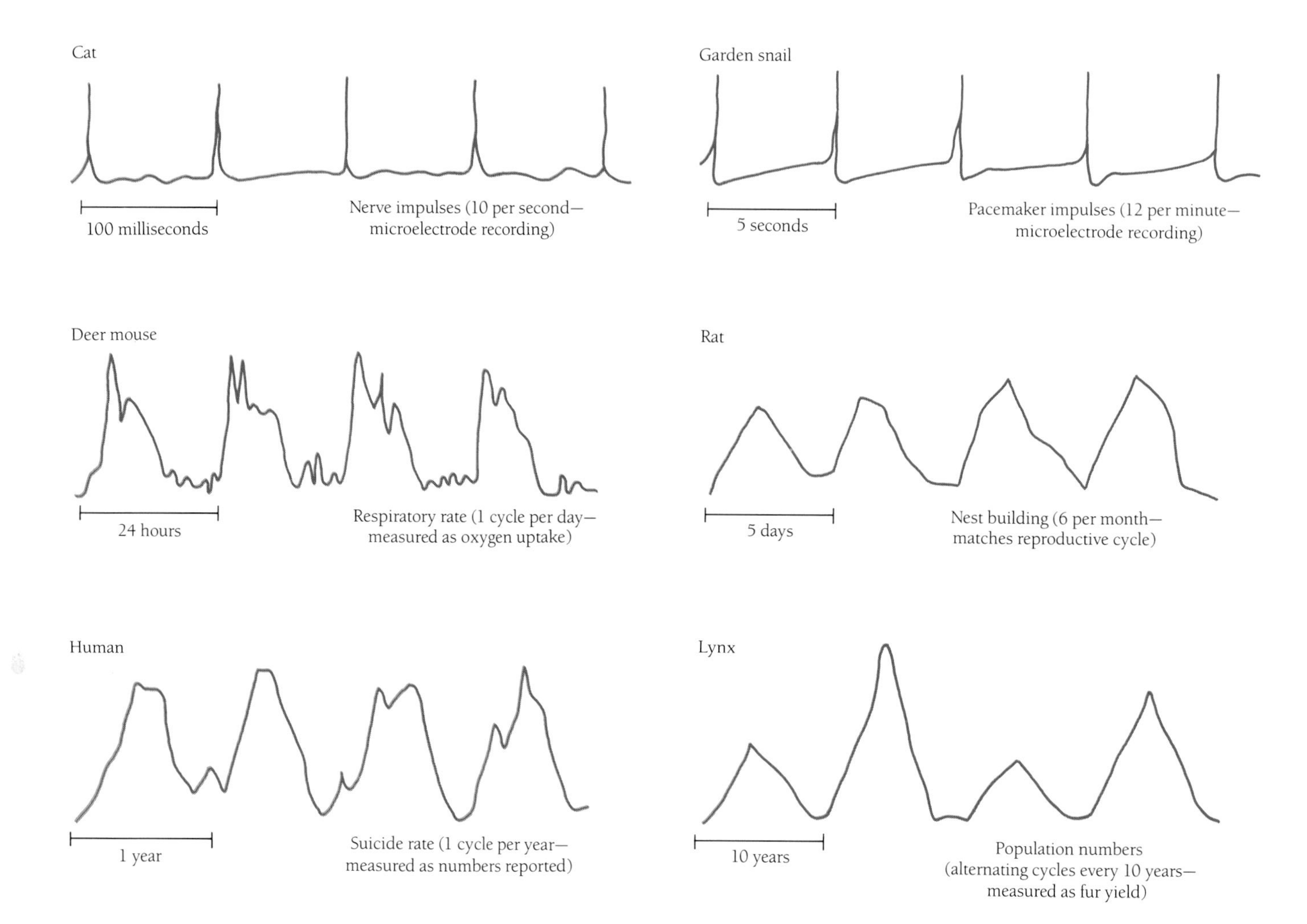

Periodic rhythms characterize many aspects of animals' lives. (*Top left*) When stimulated, the cat's nerve cell fires every 100 milliseconds. (*Bottom right*) The entire lynx population of the American far north undergoes 10-year oscillations in its size.

changing levels of an animal's activity—sleeping, feeding, moving, reproducing, metabolizing, and producing enzymes and hormones, for example—are well coordinated with environmental rhythms. But the key question is whether the animal's schedule is driven by external cues, such as sunrise or sunset, or is instead dependent somehow on internal timers that themselves generate the observed biological rhythms. Almost universally, biologists accept the idea that all eukaryotes have internal clocks. (Eukaryotes are organisms that have nuclei in their cells and include most forms of life except bacteria and so-called blue-green algae. They range from protozoans to humans, from single-celled green algae to trees.) Biologists learned that

Red crabs *Gecarcoidea natalis* on their annual breeding migration, which is tied to both seasonal and lunar cycles. The crabs live in the inland forests of Christmas Island but return to the sea to breed. The migration starts within 3 days of the last quarter of the moon at the end of the rainy season in late spring.

animals have internal clocks by isolating them completely from external periodic cues. For instance apparently normal daily periods were maintained over about a week by the fungus *Neurospora* when it was intentionally isolated from all geophysical timing cues while being orbited in a space shuttle. The continuation of biological rhythms in an organism without external cues attests to its having an internal clock.

Daily rhythms

When crayfish are kept continuously in the dark even for 4 to 5 months, as was shown long ago, their compound eyes continue to adjust on a daily schedule for daytime and nighttime vision. More recently horseshoe crabs kept in the dark continuously for a year were found to maintain a persistent rhythm of brain activity that similarly adapts their eyes daily for bright or for weak light. Like almost all daily cycles of animals deprived of environmental cues, those measured for the horseshoe crabs were not exactly 24 hours. Such a rhythm whose period is approximately—but not exactly—daily is called *circadian*. For different individual horseshoe crabs the period ranged from 22.2 to 25.5 hours. A particular animal typically maintains its own characteristic cycle duration with great precision for many days. Indeed stability of the biological clock's period is one of its major features even when subjected to considerable changes in factors, such as temperature, that would be expected to affect biological activity strongly. Further evidence for persistent internal rhythms appears when the usual external cycles are shifted—either experimentally or by rapid east-west travel over great distances. A corresponding displacement (like our "jet lag") in the animal's normal daily program typically results. This generally affects the setting, or phase, of the biological clock but not its period. Disorienting effects of a mismatch between external time cues and internal schedules may persist for several days or weeks until certain cues, like those of the day-night cycle, reset the organism's clock to the new local time.

Setting the rhythms

Animals need natural periodic signals like sunrise to maintain a practical cycle whose period is precisely 24 hours. Such an external cue not only adjusts the animal's day to begin with the local solar day but also—because it normally does so day after day—seems to keep the internal clock's period close to that of the earth's rotation. If the clock's inherent period is less than 24 hours, the outside signal serves to delay the start of the next cycle. If the clock's period is more than 24 hours, the synchronizer tends to accelerate the new cycle's beginning. Yet despite this effective resetting of its phase, the animal's timer itself continues to have its own genetically built-in period close to, but different from, 24 hours. Without the external cue, the approximate biological "day," like the tides, drifts continuously in relation to the solar day. Such forward or backward phase slippage has been extensively studied in many animals and in biological activities ranging from hatching of fruit fly eggs to wheel running by squirrels. Light has a predominant influence in setting the clock. Even a 15-minute burst of light in otherwise sustained darkness can reset an animal's circadian rhythm. Normally internal rhythms are kept in step by regular environmental cycles. For instance, we have already seen that if a homing pigeon is to navigate with its sun compass, its clock must be properly set by cues provided by the day-night cycle.

When the sun comes up, this pearl crescent moth on which the dew has settled remains inactive until the air warms. For moths of this species (*Phyciodes travos*), the daily periods of activity and inactivity are timed by the temperature cycle.

In addition to light other environmental cues may synchronize animal clocks. In some cases pulses of darkness that interrupt otherwise sustained light reset clocks—typically in the opposite direction to resetting by light. Thus if light during the dark phase sets the clock ahead, darkness during the light phase sets it back. Temperature changes of about 1° may also act to entrain the rhythm of a cold-blooded animal like a lizard. Rising temperatures are like increasing light, in signaling the day phase of the cycle whereas decreasing temperatures signal the night phase. Salinity as well as temperature have clock-setting effects on shore-living crustaceans such as crabs.

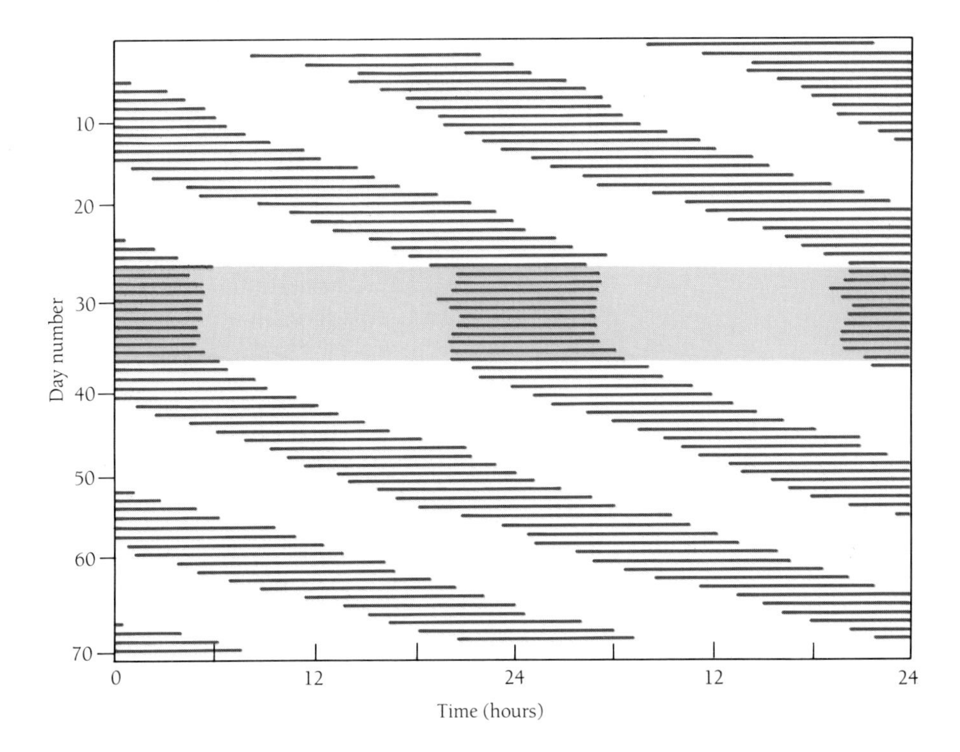

The sensitivity of a scorpion's eye changes twice daily in a circadian rhythm, reaching a maximum for perceiving light at night and a minimum during the day. The horizontal bars indicate the times of greater sensitivity; the spaces between the bars represent the times of lesser sensitivity. The record for each day has been plotted twice on the same line to make the pattern clear. For the first 26 days a scorpion was kept in constant darkness, during which the rhythm continued spontaneously but had a period of 25.2 hours. The cycle drifted 1.2 hours each day. On days 27 to 37 (green band) the animal was exposed to periodic light in imitation of normal 24-hour day-night cycles. During this time the eyes' rhythm closely matched the 24-hour solar cycle. For the remaining days the scorpion was again kept in constant darkness, and the rhythm resumed its 25.2-hour period.

The regulation of biological clocks by light is similar in invertebrates, in vertebrates, plants, and single-celled organisms. Do all these living things share some common, extraordinarily general mechanism for this function? In most—but not all—vertebrates light-sensitive elements outside the eyes aid this clock regulation. In fish, amphibians, reptiles, and birds, the pineal organ and in some species other parts of the brain like the parapineal and hypothalmus serve in this way. But in *mammals* the eyes alone fulfill this function. The input from the eye to the brain in these animals includes not only the visual pathway but also a parallel nerve tract ending in a part of the forebrain, important in regulating the clock.

Daily rhythms have been intensively studied in animals from protozoans to humans. Virtually all physiological or biochemical functions in living things fluctuate regularly on a daily cycle. Twenty-four hour patterns of oriented movement are examples closely related to navigation. For instance, we have seen that birds and bats may roost in one place and feed in another separated by considerable distances over which they migrate every day. Such tropical marine species as parrot fish that feed over open sand flats and retire for the night to the shelter of nearby coral reefs follow a similar regime of daily movements.

Daily vertical migrations

Many aquatic animals—both freshwater and marine—migrate upward and downward in the water each day. Much of the zooplankton of seas and lakes swims toward the water surface during the late afternoon and early evening and moves again into deeper water during the early morning. Vigorous swimmers like some squid and fish also migrate vertically every day either on their own or in following prey that behave rhythmically in this way. The distances traveled in these daily vertical movements are short compared with many horizontal global migrations. The deepest place in the oceans is less than 12 kilometers from the surface and the longest daily vertical migrations by animals are probably only a fraction of this, usually not more than 500 to 1000 meters. On the other hand the biomass that moves up and down every day is enormous. The habitats through which these migrators move vary greatly over short distances. In swimming from the surface to 1000-meter depth, say off Bermuda, an organism moves from a region that is warm, relatively productive, brightly sunlit, and wave agitated to a much cooler, quieter, and more sparsely populated one with little or no daylight at all even at noon on a clear day. The pressure at 1000 meters is about 100 times that at the water surface.

A number of intriguing questions about the navigation of daily vertical migrations still remain largely unanswered. How is the timing of such daily activity regulated, for example? Because all eukaryotes seem to have biological clocks, we might expect that internal rhythms could cue this plankton behavior. Indeed, some laboratory observations have shown that periodic vertical movements may continue in constant darkness. On the other hand, acoustic sounding in the sea and laboratory studies both show that an animal's depth in the water is strongly influenced by the light intensity around it.

Oceanographers have found, for example, a remarkable correspondence at dusk between the changing depth to which a given light intensity penetrated and the depth maintained by planktonic animals in the water. This suggests that the animals swim upward when the ambient light intensity drops below a given level in the late afternoon and downward in the morning when brightness becomes greater than the reference intensity. Recent experiments have shown that such behavior in the arrow worm *Sagitta crassa* is not oriented by the direction of light but probably activated by its level. In any case, because the migrating animals form a layer about 25 to 45 meters thick, the range of light intensities in this mobile population must be considerable. Furthermore even the most actively migrating species have been found over a thousandfold range of illumination. Even though light intensity must play an important role in vertical migratory behavior, other factors, including biological clocks, are certainly involved. Biologists found long ago that crustaceans occurring at 200 to 400 meters depth at night start down toward their midday levels between 800 and 1000 meters long before there is significant daylight even at the surface, which suggests that their daily migrations are timed internally.

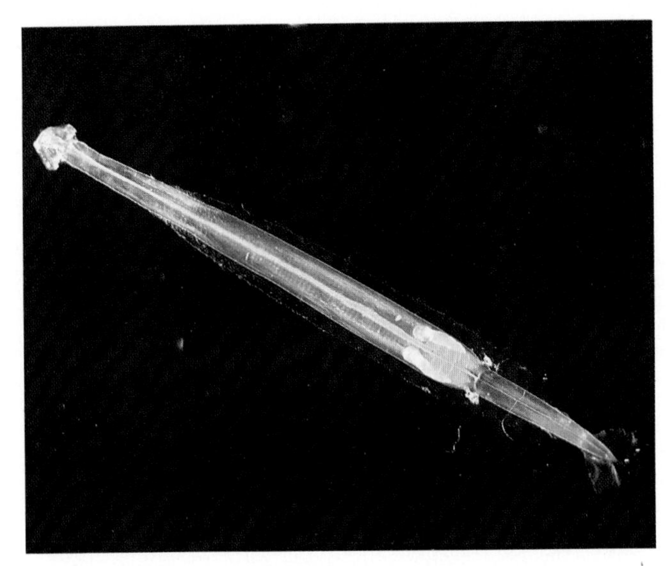

Vertical migrators found near the surface at night and in deeper waters during the day. (*Left*) The lantern fish *Myctophus asperum*. (*Right*) The small arrow worm, which is part of the zooplankton.

Tidal and lunar rhythms

Many animals, marine species in particular, have strong rhythms clearly related to tidal and lunar periodicity. Tides differ in various parts of the world even though the moon (the lunar month is 29.5 days) commonly has marked effects on both their period and height. For instance, tides may be semidiurnal as is typical of the Atlantic coast of western Europe and eastern North America where there are two tidal cycles a day with a 12.4 hour period. Here the phases of the moon influence the height of tides. Along the shores of the Gulf of Mexico and in Southeast Asia there is only one tidal cycle a day, with a 24.8 hour period, and tidal heights do not depend closely on phases of the moon. On most shores of the Pacific Ocean both tidal heights and intervals between successive phases vary from day to day.

Intertidal animals of many kinds must have rhythms corresponding to tidal cycles as must numerous other species that feed on exposed shores at low tide or time their migrations to use or avoid tidal currents. Internal rhythms timing such behaviors have been demonstrated in quite a number of animals. They usually are not as strong as daily rhythms nor are they as persistent under constant laboratory conditions. In the green crab internal tidal rhythms can be reset by temperature, pressure, or salinity, which in the sea usually cycle along with the tides. Other animals may have weak internal rhythms, or none at all, that match tidal or lunar periods even though their natural behavior shows such cycles. In such cases external cues must drive the biological rhythms.

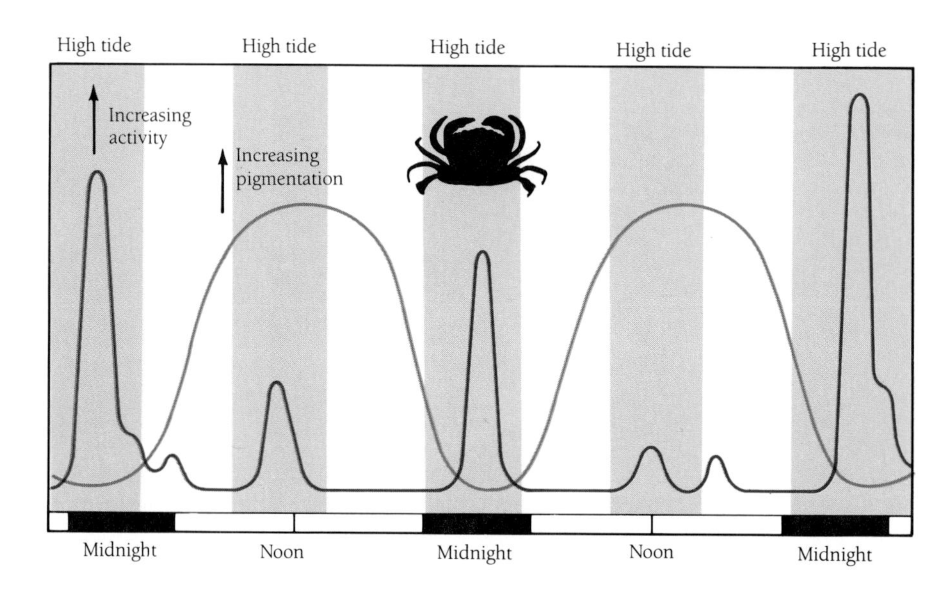

The green crab *Carcinus maenas* exhibits both tidal and daily rhythms. The crab walks about and feeds during high tides, but these activities are much more pronounced during the nighttime high tides than they are during daytime high tides. A second function, coloration, is controlled by a daily rhythm alone. During each daily cycle, the crab's pigment cells expand, making the animal darkest at midday, and then contract until it becomes palest at midnight. The tidal rhythm signals the best time to feed, whereas the color changes make it inconspicuous.

By influencing tidal cycles, the moon strongly affects animal rhythms. Moonlight also has more direct effects on certain animals. At full moon, some nocturnal insects, birds, and mammals increase or reduce their activity. Predators may be more active in bright moonlight because they can see to hunt better; prey made more vulnerable by light may be less active when the moon is full. At this time, too, plankton often shun the surface layers at night and migrate less far upward than they do at darker times of the lunar month. Thus plankton from day-depths of 400 to 500 meters, which may actually reach the surface at night during other lunar phases, are almost never found there at full moon. Both the Pacific and Atlantic palolo worms year after year time their spectacular swarming and breeding behavior at the sea surface very closely with certain lunar months. A related species from the Adriatic Sea controls reproduction with an internal clock running on an approximately lunar cycle. Lacking the precision of circadian timers, this clock can nevertheless run under constant conditions and be synchronized by a 4-day sequence of moonlight pulses! Many corals, the horseshoe crab, and the grunion (an American Pacific coast fish), as well as other marine animals spawn only during certain parts of the lunar cycle, but the relation between such reproductive behavior and clock mechanisms needs further research.

Seasonal rhythms

Seasonal cycles govern various functions in many animals, regulating not only migrations but also reproduction, hibernation, and molting. In many species whose individuals live more than 1 year, internal, approximately annual rhythms govern

The grunion's spawning is governed by a combination of tidal, daily, lunar, and seasonal rhythms. This Pacific fish (*Levresthes tenuis*) spawns at night during spring tides in March to August. "Spring" means exceptionally high flood and exceptionally low ebb tides that occur several days after both the new and full moon. Coming up onto the beach at the peak of these very high tides, the females dig their tails into the wet sand and deposit a few thousand eggs, which are fertilized as the males embrace the females. The adults return to the water by flinging themselves into a receding wave. During the next high spring tide, 2 weeks later, the eggs hatch, and the larvae are washed out to sea.

the regular period of such functions. Insects, mollusks, reptiles, birds, mammals, and coelenterates such as jellyfish have such yearly rhythms. Annual changes in photoperiod—the proportion of daylight in a 24-hour day—are a major factor in synchronizing the biological yearly rhythm of many animals with the seasons. In some species the changing light-dark ratio seems to drive the annual period directly. Yet for others an internal biological clock plays a part. For instance garden warblers kept under conditions of constant day length and temperature maintained a rhythm that was approximately annual for more than 3 years. Similar research on other warblers demonstrated internal rhythms over even longer periods. The cycle that persists in warblers is only 10 months long and therefore like internal circadian rhythms needs an external signal to set it to 12 months. In sum biological rhythms with periods ranging from a day to a year invoke a whole hierarchy of timekeeping systems. Some of these quite obviously have an internal beat while others apparently do not.

For still longer time intervals the seemingly random variability in the year to year catches of commercial fisheries have always been a puzzle. Recent analysis suggests, however, that fluctuating salmon catches and herring population changes may be correlated with slow oscillations of sea level, temperature, and salinity hav-

Every fall amateur ornithologists come to Hawk Mountain, Pennsylvania, to observe the migrating hawks and eagles flying over the ridge. Data collected by these people over a 14-year period show that each species has its own earliest peak and latest times for passing through this air space.

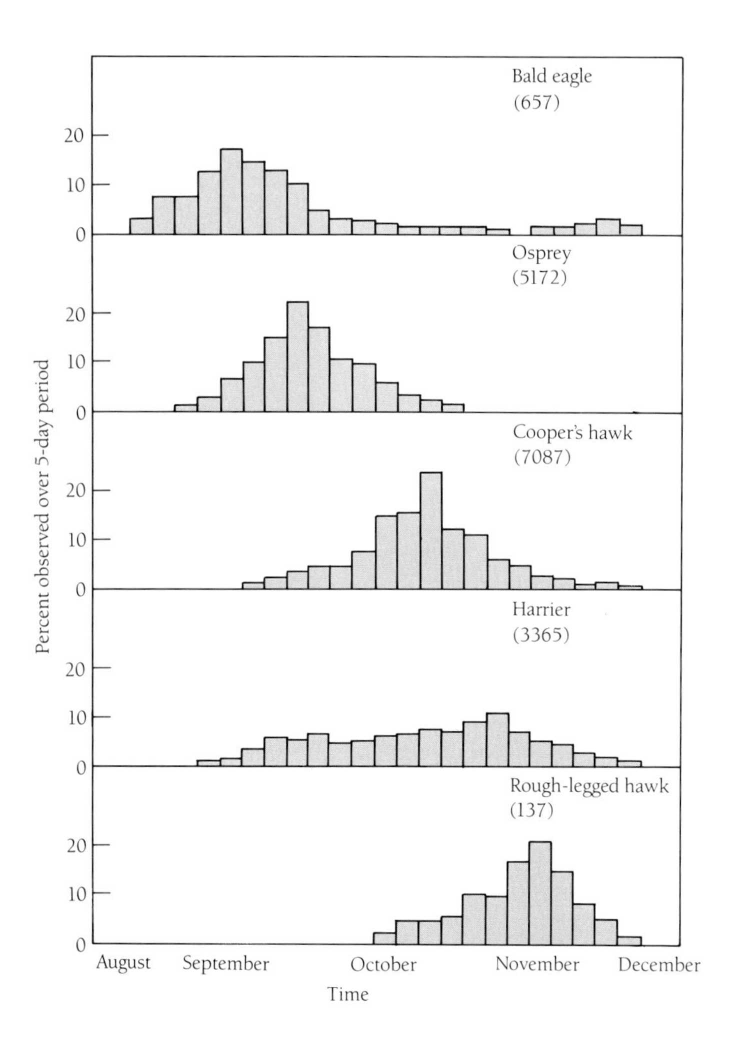

ing a 5- to 6-year period. Studying biological rhythms with such interannual periods is clearly a task for the dauntless. In addition, the problems of controls and "constant" conditions are taxing in the study of any of these animal cycles.

KEEPING TIME

However, to return to a more direct consideration of the animal clock and calendar, there is evidence in several well-studied species (mites, insects, birds, and mammals) that seasonal and annual timing may be paced by the internal daily clock as pro-

The change in the garden warbler's migrating direction is timed by an internal clock. (*Left*) In late summer and early fall the birds fly southwest over the Straits of Gibraltar to northwest Africa. Continuing the journey in the late fall, the warblers change their heading to the southwest in order to reach winter quarters in central and southern Africa. (*Right*) The seasonal change in course is mimicked by the migratory restlessness of caged warblers exposed to a constant day length and temperature. The arrow in the upper circle points to the average direction in which the restless birds orient themselves in the early fall; the lower arrow shows the average late fall direction.

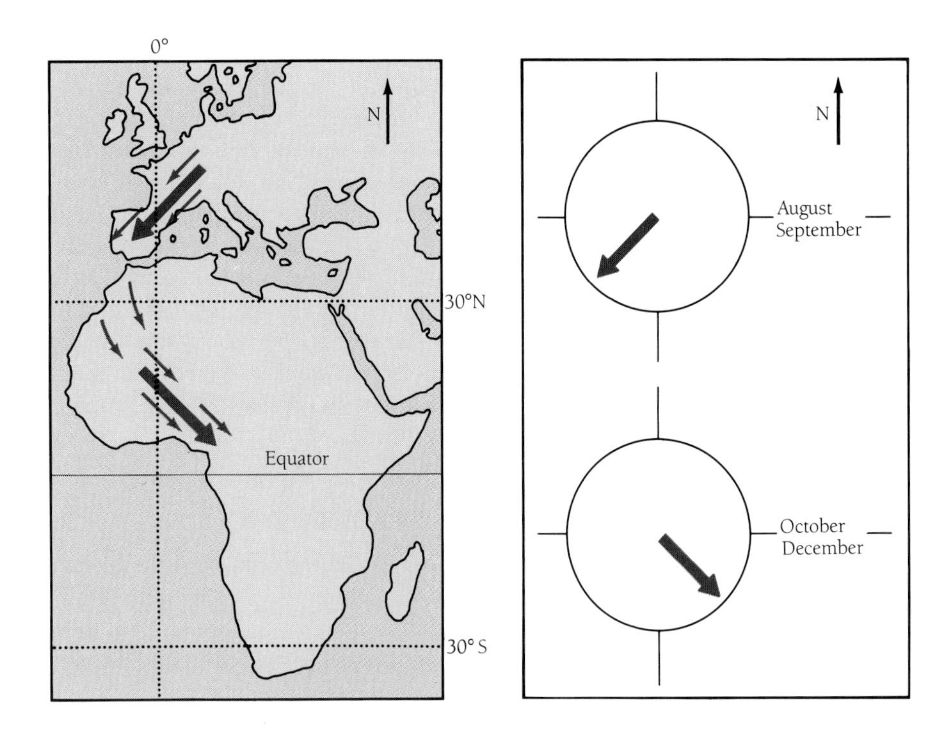

posed more than 50 years ago for plant rhythms. If so, circadian cycles must be counted to measure the longer periods. This could be analogous to the quartz crystal in a watch whose high-frequency basic rate controls not only the second, minute, and hour hands but also the calendar and even the rate of the stopwatch. In this connection a stopwatch, rather like an hourglass, can measure the duration of any event or its components but cannot indicate ongoing local or solar time. A number of behavioral, endocrine, and other biological cycles seem to act in this fashion. In such a case the time measured begins when the stopwatch is started or the glass is turned over, and the preset duration ends when the watch sweep stops or the sand runs out. For instance, most animal eggs, including the human ovum, do not start to develop until they are penetrated by the male sperm cell. Thus the timed and orderly generation cycle begins only when this partly fortuitous event triggers the system.

With regard to navigation, time-interval measurement is required when rate of movement or drift as well as distance traveled at a given speed, must be estimated. Obviously in dead reckoning and vector integration such stopwatch information is crucial. Precise measurements of time intervals are also essential for the echolocation of porpoises and bats, as well as sound localization, say by the barn owl, in which very short, millisecond periods must be sensed. Yet when the animal's time must be matched to some kind of environmental cycle like the solar day, an internal continu-

A display of chronometers. These instruments, so important to modern human navigation, are simply precise clocks.

ing timer is required and as we have seen, its phase must be synchronized with the cycling of periodic events in nature.

Scheduling of all daily activities like waking, sleeping, feeding, and migration could be regulated on this basis. Using the von Frisch conditioned feeding technique, honey bees, for instance, have clearly been proved to have a sense of local time. If a particular experimental feeding station contains sugar water only between 11:00 and 11:30 A.M., workers readily learn to visit it only around this period. Such timed control of behavior, as well as of reproduction and social activities in general, requires that the temporal sense of the various individuals coincide; otherwise all attempts at cooperation would usually fail. No doubt the sun may be read in some cases like a literal sundial to tell time. Longer time intervals such as days or monthly lunar phases must also be commonly counted in some way to cover extended activity periods.

Nature of the clock

What is the mechanism of the biological clock? How and where does it tick? Any clock requires an ongoing repetitive process that is uniform and can be counted as a cumulative series of cycles from the beginning. The swinging of the driven pendulum in a mechanical clock and the oscillations of cesium atoms in an electric field in an atomic clock are practical humanmade examples. Such a steady oscillator, which alternates periodically between two mechanical, electronic, or chemical states, is thus one of three essential elements of a timer. This pacemaker, which generates a train of countable uniform time periods, can be considered the heart of the clock itself. The second key element is the setting mechanism, which permits the pacemaker counts to be kept in phase with environmental time. Such phase-setting by the action of light on photoreceptors has already been cited in some detail. The output components of the biological clock are its third essential part and express the rhythms as they are usually observed, for instance, as sleep-wake cycles or seasonal migrations.

The clock of the sea hare *Aplysia californica,* a well-studied example, is centered in a small network of interacting neurons near each eye. Even when isolated in tissue culture, an eye of this species maintains a circadian rhythm of nerve discharges for days under constant conditions and can be phase shifted by light just as in the intact organism. The eye's sensitivity to light is important in such phase shifting, which may be mediated through protein synthesis and chemical messengers. Special fibers in the sea hare's optic nerve transmit a circadian rhythm of neural impulses from the clock to the brain. From there the nerve signals are widely distributed throughout the central nervous system to various localized centers on both sides of the body. The underlying circadian rhythm is thus available everywhere in the animal to influence all its specific daily rhythmic functions.

A relative of the snail, the sea hare *Aplysia* retains only a small remnant of a shell, which is buried in the soft tissue that covers the animal. The finlike projections are used for swimming.

This system of generating and distributing clock information in the sea hare seems closely similar in principle (but obviously very different in frequency and other details) to the pacemaker in our heart. A small node of specialized cardiac muscle originates the regular beat, which is conducted over the whole heart to trigger and unify the muscular pump's contraction. Actually the heart analogy may be interesting for a further comparison. Both *Aplysia* and our heart apparently contain many oscillators that are normally integrated into a single periodic mechanism. In vertebrates all parts of the heart muscle are themselves rhythmically contractile. It has been known for a long time that even individual embryonic heart cells will beat on their own in tissue culture. However, a specialized nodal pacemaker ordinarily acts as the *master timer* initiating the whole organ's normal beat. Overall daily rhythms in many animals also seem to depend on a number of cells or elements with their own periodicity. Each timer potentially controls some specific cycle like the synthesis of a certain hormone or regulation of body temperature. Yet normally these are all coordinated in an organismal daily pattern. The existence of such different timers only becomes clear under certain experimental or abnormal conditions when one or more may break away, each with its own phase and period.

Where is the clock?

Does the eye-related clock of *Aplysia californica* provide a model for other biological timers? Where are they and how are they synchronized with external cues? In a variety of animals both nervous and endocrine elements function in timing. In mammals a part of the brain called the hypothalamus regulates endocrine organs such as the pituitary gland. Within the hypothalamus lie paired regions called suprachiasmatic nuclei. Neurobiologists think circadian rhythms are maintained here because rats and monkeys stopped expressing circadian rhythms after their suprachiasmatic nuclei were surgically removed. Because the suprachiasmatic nuclei receive input from the retina, they may also be sites where rhythms are synchronized with external cues. Direct sensitivity of the pineal gland of birds, fish, amphibians, and reptiles to light is also important because it allows environmental signals to synchronize the animals' spontaneous circadian rhythm with the day-night cycle. Removal of the pineal results in locomotor and thermal arrhythmia in the sparrow. Such a bird acquires the donor rhythm if the diurnal oscillator, here the pineal, of another normal individual is implanted! The hormone melatonin, secreted only at night, and an enzyme important in melatonin synthesis are known molecular parts of this avian clock. In birds, then, the pineal gland apparently is the primary pacemaker; it plays the major role both in generating and in regulating their circadian rhythms. In mammals a pineal rhythm is present but is secondary and driven by neural signals from the dominant suprachiasmatic nuclei.

The clocks of various insects also reside in endocrine structures and parts of their optic tracts and brains. Biologists have proved this again by transplanting functional clocks. For instance, an arrhythmic cockroach, cricket, fruit fly, or honey bee can be induced to behave rhythmically by implanting parts of the brain or optic lobe from a normal animal. Hence the donor tissue's master clock appears to take over driving the host's lower level oscillators that contribute to its overall rhythms. In crayfish and other decapod crustaceans, neurohormones produced in the eyestalks and in parts of the central nervous systems express a circadian rhythm, but whether they are part of the master oscillators or merely driven by them is not yet certain.

Long-period controls

How do animals time cycles that last for a year or longer? The endocrine system in vertebrates helps regulate such slow rhythms. Several parts of this intricate network including the pituitary, the pineal, and the hypothalamus, provide both programming and control for migration. The hormones prolactin and corticosterone, for instance, affect migratory behavior in fish like the salmon because they function in maintaining the mineral and water balance critical for the fish's transition between fresh water and the sea; increased thyroid activity induces the maturation that precedes the salmon's downstream migration. The increase of prolactin in the late summer in the white-throated sparrow *Zonotrichia albicollis* stimulates both the premigratory storage of fat needed for the long journey and the southward-directed migratory restlessness. Whereas cues from the photoperiod govern the timing of annual cycles in birds and mammals, temperature plays a considerable role in timing

A swarm of ladybird beetles *Hippodamia* in California. As the temperature drops in the late fall, the beetles, which have been summering in the hills, congregate in very large clusters to overwinter. The cluster breaks up in the early spring when the beetles migrate down into the valleys, where the temperature is warmer. Here the population remains until hotter temperatures trigger a migration back up into the cooler hills.

the migratory behavior of a cold-blooded killifish. Genetic, photoperiodic, and hormonal factors closely time insect life cycles, including their migrations.

There may be a sort of competition between the induction of downstream migration in juvenile salmon and the early development of reproductive maturity. If so, in landlocked stocks maturing early has "won out" over the urge to travel. The seagoing migration of North Atlantic eels coincides with their sexual maturation and their metamorphosis into the silver eel. The laboratory study of eels brought artificially to maturity with pituitary extracts and gonadotrophic hormones may help explain the control systems that time this spectacular migrator. In populations of various fish developmental processes lasting several years appear as seasonal events every year, but for individual fish they obviously are not. In Pacific salmon and in Atlantic eels, for instance, the full migratory cycle occurs only once in an individual's lifetime. Even though separated by 2 to 5 or more years, the beginning and the end of such migrations are usually locally timed to coincide with daily and seasonal environmental events. Many other details of life cycles, as we have seen, are also coupled with clock and calendar. Yet the overall course of development, including inherent behavior, is clearly programmed genetically for each animal. How are such ad hoc and inherited aspects of life integrated? Although this nature-nurture competition has come up here in relation to how animals time their behavior, it applies to our overall topic as well. Indeed its proper discussion has been saved for the last chapter, which we may now begin.

10

Why Animals Migrate

So far we have considered the goals of migration as if they are something already chosen like the set point on a thermostat. Now we need to ask: What sets or selects an animal's goals? As the locomotor and navigational abilities of long-range migrators can take them almost anywhere on earth, why do they have particular, often precise, goals, or series of them? To answer this question we must consider the animals' basic dependence on their environments and on their biological histories. Hence we need to ask some questions about the ecology and evolution of these navigators. How can the relations between animals and their environment make migration worthwhile? How have massive changes in the surface of the earth itself as well as the variety and vigor of biological evolution affected migration?

ADAPTATION AND MIGRATION

Wandering albatross nesting. With its very long, narrow wings like those of a sail plane, the albatross glides downwind on its continuous migration around the antarctic continent. The bird seems to be a nocturnal feeder that eats squid and other pelagic animals that move up to the sea's surface at night. Each circumpolar trip takes about a year, and the migration is interrupted every 2 years when the albatross breeds on one of the remote antarctic islands like South Georgia in the South Atlantic and Kerguelen in the Indian Ocean. Banding records show a high degree of fidelity to the nesting site.

An animal's home territory must satisfy all its primary needs. It must have appropriate physical and chemical features such as temperature, water, various minerals, and oxygen. It must also contain the right kinds and amounts of food to provide the energy necessary for survival, growth, and reproduction. Mates must be available to reproduce sexually. How well the organism can select a habitat that has all the required resources, as well as use and if need be, defend them, is a measure of its *adaptedness*. Avoiding such negative features as competitors, predators, and parasites is also part of successful habitat selection. Because the world offers a far from uniform environment, choosing a proper place to live requires that animals discover what options are available and select the best. Hence exploration for an optimal habitat, the effective use of its resources, and avoiding any dangers present all require an ability to move about and develop a familiar territory. For instance, both

Honey bee foraging among apple blossoms.

animals and their food sources typically are discrete localized objects scattered in a geographically patchy environment. As a result active browsing or hunting is a central aspect of most behavior; such foraging is also a key feature of behavioral ecology. Because most of the other resources needed or dangers to be avoided are not uniformly present in the habitat, locomotion and navigation are also needed to find or avoid many other environmental features. Thus the patchiness of the world in time as well as space, and the complex needs of individual animals provide the potential benefits of migration and thus navigation.

Adaptation fits an animal to its environment like a hand to a glove. This fit must exist at many levels that vary in complexity and time scale. Sometimes animals relate to the environment by using behavioral reflexes, like evading a threatening predator or courting an attractive mate. If so, they respond immediately to particular stimuli in performing such short-term, nearly automatic acts. In other cases when learning and memory are required, a longer time scale is necessary to develop an adaptive response, which may far outlast an immediate reflex. Thus one or more encounters with a predator in a certain place may cause an individual thereafter to detour around it or be more alert when there. Learned behavior lasts as long as its memory trace that may persist from seconds to years. Weakly social animals can transfer such behavior from one to another by imitation and reinforce it by repetition. Strongly social species can share learned adaptive responses far more complex than simple sensory reflexes by communicating as honey bees do, or by teaching, as does a mother bear with her cub.

On a much longer time scale, the possibility emerges that both individual and group learning, through natural selection, can be inherited and influence the course of evolution. At the population level the survival value of particular behavior extends to future generations of the family, the group, and the whole species. Hence we should consider here not only the nature of heredity but also its ecological and evolutionary influences on animal navigation and migration. Obviously this is such a large topic that only a few implications can be outlined here.

Meeting adversity

Animals have several options to minimize the harmful effects of an unfavorable season. Even so, all of these may involve greater hazards than those expected during the best time of year. One alternative is to store seeds or other food when it is abundant for use when it is scarce. Many birds, mammals, and insects do this. Another possibility is to go into a state of suspended animation, such as *diapause* (typical of many insects), *hibernation* during winter cold, or *estivation* during summer heat and drought. These are all conditions in which the animal greatly reduces its activity and metabolic rate so that it requires little or no food. Moreover it may develop remarkable resistance to freezing temperatures or other conditions that would be lethal in the normal active state. Certain small mammals, a few fish such as

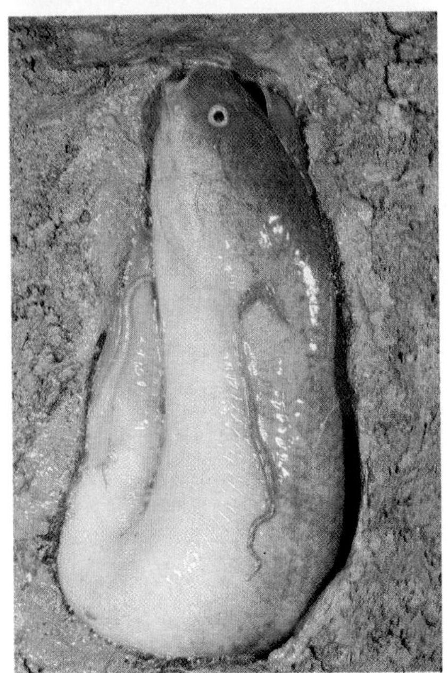

(*Top*) A cluster of hibernating Indiana bats *Myotis sodalis*. (*Bottom*) The lungfish *Protopterus annectens* curled up in a mucous cocoon under the mud is in estivation, a response that allows the fish to survive the seasonal drying up of its pond.

lungfish, some amphibians and many insects as well as other invertebrates "cop out" in this special manner.

Of course, the third alternative is to migrate, moving to a place where conditions are more favorable. Periodic cyclic movement between parts of their range is a way of life for many animals. A migratory habit could have arisen through exploration that led to a favorable discovery or through an attempt to escape temporary adversity. If migration significantly improves an animal's habitat, it may be expected to pay off with reproductive success. In support of this notion the great majority of commercially important fish are vigorous migrators. It surely seems plausible that traveling from one productive feeding area to another contributes to their enormous numbers. Spectacularly dense populations of other animals, too, may well result from their migratory habits—huge bird flocks or butterfly and locust swarms, for instance. That population growth depends on the resources available seems quite reasonable. Yet as with the proverbial hen-and-egg paradox one could also argue it the other way. In which case migration, or at least emigration, would be required for survival when any species reproduced beyond the resources present in its native haunts. That crowding is a key factor in inducing the migratory phase of the desert locust would fit in well with that alternative. The occasional occurrence of one-way expansion could lead to a migratory cycle should that become adaptive.

Costs and benefits

The question of "why navigate?" can be answered in an interesting way from a cost-benefit point of view. In this sense navigation is a way of critically reducing the effort needed to move from place to place by making it cost-effective. Migratory trips that would not "pay off" if attempted in a casual, bungling, or inefficient manner could clearly become advantageous provided that skillful navigation minimized the energy, dangers, and time required. Usually the ecological or evolutionary success of any species is gauged by its ability to increase in number, or at least replace itself from one generation to the next. Some ecologists try to measure or estimate the costs and benefits to animals of various behavioral alternatives in terms of their effect on survival and reproduction. In general, migration increases the quantity or quality of resources potentially available to an animal. But whether migration is advantageous for a particular species depends on such factors as the effects of other species that may be competing with it or preying on it. For instance, in the African Serengeti fewer nonmigrating hoofed animals like impala and warthogs are killed by predators than die of starvation because of fierce competition for severely limited food during the dry season. But large carnivores such as lions are the major lethal factor for long-range migrators like Thomson's gazelle and zebras. Similarly the benefits of periodic movements may depend not only on population density but also markedly on habitat stability or its regular seasonal variation. In one model investigating the

Migration of wildebeest *Connochaetes taurinus*, a hoofed mammal of the African plains.

reproductive benefit of an animal's moving from one part of the habitat to another, the basic test compares migrators and nonmigrators—movers and stayers.

The migrators must invest energy in locomotion and in some cases be exposed to more dangers en route than at home. The model compares these losses with the mortality of nonmigrators under the poor conditions of the off-season and also with the gains achieved, for example, by the migrator through their having optimal forage all year. On the other hand stayers may benefit from being first in position to breed and from their uninterrupted site familiarity, but may face as penalties shortage of food, severe weather, and life-threatening temperatures. Conversely, brief but great summer productivity at high latitudes allows these regions temporarily to support large immigrant breeding populations. This hospitable abundance may be critical for the successful reproduction and rearing of young by many migrant species.

Note however, that the diversity of behaviors and situations we have encountered in our review of animal migration offers many challenges to cost-benefit analysis. Neither the salmon nor the eel feed at all in their terminal breeding areas, for example. Some baleen whales also apparently have comparable long periods of starvation in their temporary warm-water reproductive zones. In such instances adequate energy both for reproduction and for subsequent migration must be saved up in advance. In general the acquisition and storage of the necessary fuel are important prerequisites for migration even on the small scale of honey bee foraging. A worker departing the hive typically stocks up with the right amount of honey for its foraging round trip. Engorging too much would reduce its carrying capacity for nectar and pollen when homing. Too little could stall its return to the hive. Benefits of certain navigational behavior in one animal may even be subverted to serve the needs of another. Like ship wreckers, certain predators produce false directional signals that lure individuals of another species to offer themselves as prey! Females of one kind of firefly, for example, simulate the male-attracting sequence of light flashes normally used by females of another species. Misled males of the imitated species respond to the false signals and are devoured. In a variety of moths, such as the silk moth, courting males are drawn toward and trapped in the webs of bolas spiders by a comparable subterfuge. The spiders have evolved a counterfeit pheromone that smells like that of the female moths the males are seeking. Chemical homing on the source of this sham signal leads the insect to its death instead of to a sexual encounter. Clearly, cost-benefit bottom lines *can* become rather diffucult to estimate.

Life cycles and reproduction

Because environments usually change with season (or even time of day) and an animal's requirements are often quite different in various stages of its own life history, ecological relations typically become much more complex when considered over time. The larva of a particular insect species may, for example, be a crawling, leaf-eating wormlike caterpillar, whereas the adult of the same species is a free-flying, nectar-sipping butterfly. Clearly various other needs of these two stages differ as markedly as do their foraging habits. Many parasites have complex life cycles in which stages living in different hosts, such as fish and humans, are linked by still other stages that are free living. Similarly many marine invertebrates like worms, mollusks, and crustaceans begin life as minute free-swimming planktonic larvae that feed very differently from the adults. An adult lobster, for instance, crawls around the bottom at night to feed on mussels, worms, fish, seaweed, and other things in great variety. The minute lobster larvae feed on microscopic plankton floating in the

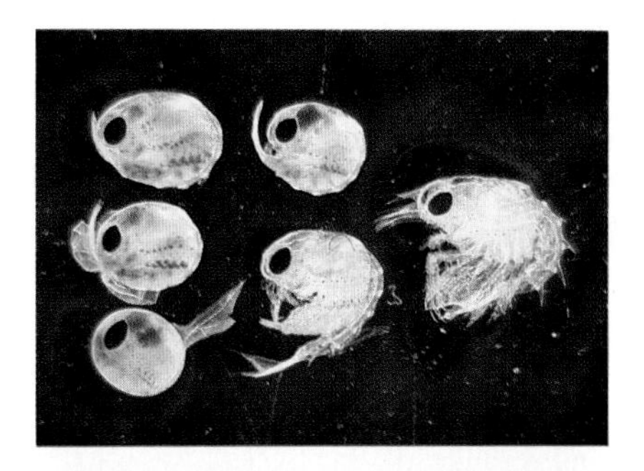

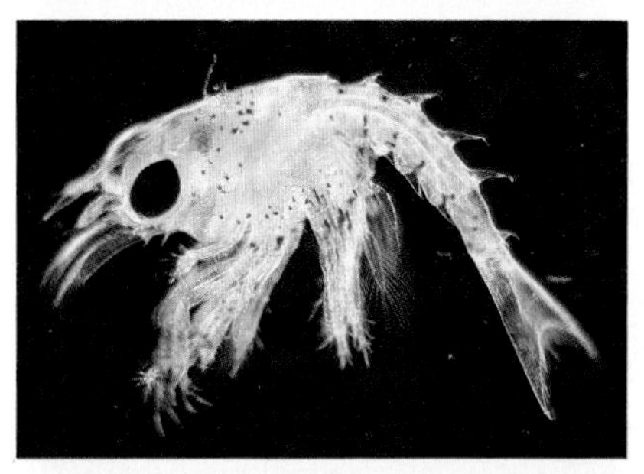

Egg and larval stages of the American lobster *Homarus americanus*.

water. The successive stages of many fish forage in their own ways and actually migrate during development from their spawning grounds, first to larval or juvenile feeding grounds, then on to those of adults. Recall that we humans as we grow and mature change our food requirements from mother's milk to baby food to adult fare. Quite different "foraging places" are typical for each of the three. Because of these factors in animals, migration and the navigation it requires may be considered as a means of properly locating successive phases of their development. Such adaptation to biological growth cycles is different from, but often in addition to, repeated migratory responses to environmental rhythms such as the seasons, the tides, and day length, or photoperiod.

It is not surprising that endocrine and other elements that regulate seasonal movements also govern development and reproduction. In fact migration can be considered as a part of an animal's life cycle. For instance, while monarch butterflies are inactive during the winter in the central Mexican mountains, they are also in a state of reproductive arrest from which they emerge in the early spring to copulate before starting to migrate north. Many birds migrate in the spring from temperate to high latitudes to breed in the summer. Gray whales have a similar pattern, but both north-south polarity and season are inverted. Growth, molting, and reproduction in birds usually alternate with migratory activities. In large waterfowl, for example, all the flight feathers are molted simultaneously so that for some weeks loons, ducks, geese, and swans are quite unable to fly until their plumage has grown in again. A number of such species specifically migrate to a safe place for molting right after reproductive duties are over, often in late summer. Smaller birds lose their flight feathers in sequence but time their molting so that it does not interfere with migration. Birds that are long-range migrators often molt twice a year in contrast to the once-a-year molting of sedentary populations or closely related nonmigratory species. Nest building, feeding, and caring for the young appear quite incompatible with extensive travel since hatchlings are helpless and immobile. The extensive expeditions of emperor penguins and gray-headed albatrosses to gather food for their chicks may seem to be exceptions. Most long-range travelers show well-timed seasonal programs that control their sedentary and migratory phases. Thus several weeks or months of physiological preparation, particularly involving buildup of fat as an energy source, may precede migration. The flight muscles of small birds about to travel typically increase in size and their body mass may double. Such metabolic foresight empowers land birds to fly far over the sea or desert where no food is available. Anticipatory timing of the ripening of reproductive organs permits various migrators to breed promptly on arrival at their spawning or nesting area. Particularly for species whose migration takes them to breeding areas where the favorable season is precariously short, this is crucial in enabling the young to be successfully reared. As we have seen, such timing depends on internal clocks and various hormonal cycles synchronized by external signals such as photoperiod.

According to one hypothesis, maturation and migration oppose each other in juvenile salmon, deflecting behavior in one direction or the other as the environment changes. When food is abundant, the alevin postpone downstream migration and adaptation to salt water. They then rapidly mature sexually. Under certain conditions there may be no seaward migration as, for example, in landlocked sockeye salmon and in rainbow trout. The precociously mature nonmigratory salmon are smaller when they reproduce than those that go to sea and feed for an extended period. The hypothesis may explain why in the St. Lawrence River, whose Atlantic salmon population has been decimated by overfishing and pollution, many individuals that are small, sexually mature, and sedentary have recently been found. If, for instance, a wild salmon population were reduced by severe overfishing at sea, natural selection might bring about some shift in its genetically determined behavior. Thus genes that conferred a strong urge to go to sea would tend to be suicidal; many fish with these genes would be eliminated. In contrast fish whose genes made them rapidly maturing and weakly migratory or nonmigratory would have a better chance to survive and would become a more obvious part of the population.

A comparison of the life cycles of the anadromous salmon and the catadromous eel reminds us that the migratory patterns of a species were determined in part by its evolutionary history. The salmon normally must go to sea to feed, grow large, and become sexually mature. The eel acquires most of its growth by feeding in fresh water and only leaves it for the ocean as the large maturing silver eel that does not feed! How does one account for this apparent contradiction? Clearly one major factor must be the specific histories of various populations, species or higher groups. The evolutionary opportunities and obstacles as well as the capacity to deal with them were no doubt unique for each species. For example, eels may have evolved in the tropics at a time when rivers offered a better place to feed and grow than oceans, which were then somehow better breeding grounds. Through the natural selection of individual genetic differences, the cumulative results of survival or failure cause one species to become very different from another.

Behavioral choices

An animal's familiar environment, as we have seen, has a number of predictable features, like night and day or winter and summer. It also has random components, at least unpredictable in detail, like weather, chance encounters with other animals or plants, and infection by parasites or diseases. Ideally an animal should have stereotyped innate behavior for responding to the dependable elements coupled with enough flexibility to cope with the unexpected or unpredictable. Actually what animals do in general is based on three major kinds of information. These determine how it can migrate as well as the rest of its behavior. Most directly, immediate sensory stimuli evoke behavioral responses that often appear as simple approach or avoidance reactions in which the relation between input and output—the stimulus and the reaction—seems quite direct. Next in immediacy is all information about previous experience that has been learned and stored in memory. This, of course, may be continually compared in the nervous system with the current sensory input. Prediction based on what has happened repeatedly in the past obviously permits smarter choices to be made about whether and how to respond in the present. Remembering is obviously important, as we have seen, for much of animal navigation. Genetic information contributes to behavior as the third information source; it underlies and complements both current sensory input and remembered experience.

GENES AND MIGRATION

An animal's inherited genetic makeup, as just mentioned, supplies a basic source of navigational information. Not only is the animal's *present* involved here but so also is its continuity with the *past* and *future*. To begin with, genes inherited from its parents obviously determine the nature of the beast and hence what it can and cannot do. The fact that it is a worm, a crab, a fish, or a bird establishes the potentialities as well as the limits of its behavior. The expression of some inherent traits depends on environmental signals like photoperiod acting via light receptors on the neuroendocrine system. The hormones then produced control gene expression in certain key cells at particular times. Such regulation in development and behavior is evident in life cycles involving quite different environmental demands such as those of the caterpillar and the butterfly. More directly the kind of animal, its size, its habitat, its locomotion, and many other genetically determined features have marked effects on its inherent navigational and migratory abilities. In addition various components of such behavior itself may be under direct genetic regulation. Yet except for salmonids and a few songbirds almost no data on the genetic control of migration are available. Even what there is relates mainly to characteristics of large populations rather than to genetic mechanisms or individual behavior.

Genetics and behavior

Genes do closely control reflex behavior. A population of wild-type fruit flies subjected to 600 generations of artificial selection demonstrated this. Geneticists chose for reproduction those flies having the most extreme behavior in turning either toward or away from the direction of gravity. The flies were mated only with others showing the same response. After about thirty generations, repeatedly chosen in this way, two new stocks have evolved. When given a choice of turning up or down at a branch point on a path where either response to gravity could be chosen, one stock tended to turn downward whereas the other turned up. Such research shows that innate, oriented behavior may be under rather direct genetic control. If turning upward were invariably lethal in nature, the gene governing that behavior would be eliminated and the other one would survive. This same kind of selective breeding for responses to light has produced stocks of fruit flies that walk towards the light and others that walk away from it. Selection for the fruit fly's response to moving stripes has been similarly studied over the years as part of extensive research on mechanisms of its navigation. Mutant fruit flies lacking certain visual nerve cells turn less in response to moving stripes than normal flies. From what we already know about the importance of this behavior for orientation in space and for course keeping in navigation, it is clear that such a mutant fly would be less likely to survive in nature than the wild type. Experiments of this kind emphasize the value of studying both the effect of a given genetic mutation on the detailed structure of the nervous system and its influence on behavior. This approach clearly offers an important way of revealing mechanisms underlying behavior. Direct genetic control has also been demonstrated for behavioral mutations ranging from specific *taste blindness* in humans, to *aggressiveness* in crickets, to *waltzing* by mice. Additional behavioral components such as color preferences, breeding time, object fixation, circadian rhythms, pheromone production, turning responses to odor and smell, foraging patterns, and other orientational and navigational functions have also been studied genetically.

For us the main point of such research is that genetically distinct strains consistently respond differently to the same environmental stimulus. Alternative reflex behaviors can put one strain in jeopardy for survival and help another to thrive. In other words genetic variation clearly provides a basis for natural selection to act on specific elements of behavior. The time-compensated sun compass of beach fleas is at least in part genetically determined. Note that to avoid predators and to follow the tide up and down their native beach, beach fleas take bearings from this compass and head perpendicularly to the shoreline, toward the safety of the water or inland as the tide rises. Obviously the correct compass heading depends not only on the sun's bearing and the time of day but also on the direction of the home shoreline, which varies for populations in different places. Beach fleas from sites with distinctly different geographic alignment were mated, and their hybrid offspring were tested. The hybrids had either intermediate sun compass orientation or, in some cases, were disoriented. For beach fleas living where the shoreline is strongly curving or rapidly varying, a fixed single direction of escape would clearly be less useful. This conclusion is supported by the irregular behavior observed in individuals from such locations.

Genetics, environment, and long-range migration

As mentioned, not much genetic research relevant to long-range migration has been done. The available data deal particularly with fish and birds. Selective breeding of salmon can have strong effects on several aspects of their offsprings' migratory behavior. Homing reliability, timing of reproduction, and growth rates have all been successfully manipulated in such studies. In this way population changes commercially favorable for fish ranching have been successfully produced. Just how the navigational map, compass, and chronometer are affected in these practical experiments remains to be discovered. Even less certain so far are the detailed mechanisms involved at the molecular biological level. Nevertheless it is quite clear that some of the key elements of migratory behavior must be under genetic control.

Natural populations of rainbow trout differ genetically in their swimming behavior in response to water currents. Trout that spawn in streams that flow into lakes inherit a tendency to swim with the current, whereas ones that spawn in outlet streams inherit a tendency to swim against it. Similarly, wild populations of both Atlantic and Pacific salmon species associated with different native rivers have distinct genetically determined biochemical and behavioral traits. Because they home from the sea rather precisely to different spawning sites, the various stocks generally do not interbreed and for this reason have drifted apart in their hereditary makeup. The genetic pattern unique to each group undoubtedly governs certain features of each migratory program as well as other traits. As a result salmon transplants between rivers are usually unsuccessful in maintaining themselves. Therefore migratory behavior is partly determined by genetic factors, partly by imprinting and chemotaxis. Experiments to improve the status of an endangered population used sperm from remaining local stock to impregnate females brought in from elsewhere. Survival and return to inshore waters of their hybrid offspring were about equal to those of controls, but ascent of home rivers was decidedly inferior to the wild stock. Genetic analysis indicates that even for a worldwide species like the skipjack tuna there are several, probably four, diverse populations with some reproductive isolation. Similar distinct geographic groups are also known for cosmopolitan whales like the humpback. Most hatchery stocks of salmon are of mixed heredity and hence tend to lack the specific migratory abilities of wild populations. If this significantly reduces the chances that fish bred in captivity will reproduce in the wild, the value of such aquaculture for increasing or even conserving natural stocks is thereby lessened.

One aspect of the genetics of salmonid fishes is quite unusual: Their ancestors millenia ago apparently underwent a doubling in chromosome numbers that has stayed with the whole group ever since. Thus instead of the typical animal diploid genetic makeup with pairs of duplicate chromosomes, this fish family has body cells with four copies of each chromosome. Such genetic multiplication has been cited as a possible factor in the striking salt water-fresh water versatility of these fish and hence in their remarkable anadromous migration. For North Atlantic eels, which incidentally are ordinary diploids, there is little evidence of genetically divergent strains within either the European or the American populations, both of which are widely distributed. Contrasting with the stream-specific salmon, the eels have large homogeneous breeding populations coupled with larval dispersal over broad areas; this produces the observed uniformity *within* each of these populations on the two sides of the North Atlantic. Yet *between* the North American and the European eels there are distinct differences presumably including hereditary factors governing their migrations.

Population differences in migratory behavior are known for land birds, mainly for certain species that breed at temperate latitudes. A number of varying behavioral patterns have been widely observed. Some of these are triggered by environmental signals; others seem innate. A migratory range that changes with particular conditions encountered from year to year is an example of variation governed by environmental signals. Temperatures, available food, and even stormy weather at critical times may induce, cancel, hasten, or delay departures or arrivals beyond those expected from the usual seasonal calendar. For instance, in western Europe lapwings often undertake substantial southerly migrations—but only in years when their normal winter habitat suffers a particularly punishing cold spell. The lapwing is commonly called the frost bird in Spain where its weather-induced movement may make it a harbinger of cold weather.

Irregular excursions like those of the lapwing may instead depend on the buildup and decline of prey populations. When rodent prey fails, the snowy owl, for instance, may migrate far south of its normal northerly range—even as far as New Haven, Connecticut where I have seen one. Instead of varying irregularly from year to year in this manner, migration may differ regularly with age and sex within a given species. Only young and female dark-eyed juncos, for instance, migrate in the fall from areas where the winter is severe. Adult males, better able to survive harsh conditions, gain reproductive advantage from overwintering in the breeding range by establishing territories there before the nesting season begins. A somewhat similar system of sex and age migratory differences of European blackbirds has also been related to adult male adaptations.

Female blackcap warbler *Sylvia atricapilla* and chicks.

Certain bird species that breed over a wide range of latitudes have geographical subpopulations some of which migrate while others do not. Individuals breeding at higher latitudes like Poland, Canada, or Tasmania actually may leapfrog over their nonmigratory fellows that nest and live all year at more moderate latitudes. Some birds of other species migrate, while others of the same kind in the same location do not. In general the proportion of individuals migrating tends to be directly related to latitude in such partially sedentary species. In the face of uncertainties a species or population that plays it both ways (some go, some stay) usually has a better chance to survive and reproduce than one entirely committed to one or the other pattern. All the western European population of blackcap warblers migrate, for instance, while only part of the African population does. The migratory restlessness of a series of north-south samples of caged blackcap warblers ranged from most intense and longest in a sample from Finland to markedly weaker and shorter in a sample from Africa. In birds from Germany, 100 percent showed the migratory urge compared with only 23 percent among those from Africa. Hand-reared German-African hybrids were intermediate to their parents both in intensity and duration of migratory restlessness. Hence the tendency to migrate in this case must be under graded genetic control. Hereditary factors induce and sustain oriented locomotor activity in migrators. Endocrine research on the partially migratory European blackbird suggests that the key inherited element that influences such behavior controls responsiveness to hormones.

EVOLUTION AND MIGRATION

The selective effect of adaptive experience over many generations is what we recognize as evolution. The underlying hereditary mechanism for evolving organisms should ideally have two kinds of components. One of these should provide stable and fixed control of functions related to rather constant predictable features of the animal's needs and environment. The other should be flexible and opportunistic in order to cope effectively with the many factors that vary in unpredictable ways. A population closely fit by its heredity to a narrow range of environmental conditions would have little adaptability of this kind. Even in a genetically versatile population, environmental changes that are too fast, too extreme, or in a direction contrary to an animal's major evolutionary "commitment," say to fresh water or large size, obviously could lead, and often have led, to extinctions. Such curbs to adaptation presumably explain why organisms including humans are not divine and do not live happily ever after. How to explain the mass disappearance of major groups, like dinosaurs, during times of great climatic change or global cataclysm has become a matter of considerable interest and debate among paleontologists. The likelihood has been considered that rates of evolution may be fastest when stresses due to environmental changes are strong but not lethal. When successive environmental cycles including the advance and retreat of ice caps go on for hundreds of generations, the kinds of biological adaptation required for survival obviously must occur at a population level rather than just a behavioral or physiological one. The latter are immeasurably quicker but of themselves have just short-term effects on individual animals. Although these generalities apply to all aspects of animals' interactions with their environments, we are particularly interested in their relevance to the origin and evolution of migration as well as navigation in various species.

Effects of global changes

We have seen that an elaborate network of factors—biological and environmental—interact to make migratory behavior a viable, major aspect of life on earth. However, an important point has not been emphasized until now, namely that all the complex relationships operational at any one time must be viewed against a background of long-term changes that constantly affect both the living world and its nonliving environment. Species and whole major groups evolve, flourish, and become extinct

Times during which six major migratory animal groups are known to have existed. Fossils first appear in abundance in rocks that were formed about 590 million years ago, at the beginning of the Paleozoic, "old life" Era (see top of scale). The two subsequent eras are the Mesozoic, meaning "middle life" but commonly called the Age of Dinosaurs, and the Cenozoic, "modern life," or Age of Mammals. Geologists divide each era into a number of periods and periods into epochs; the names in parentheses are the periods (epoch in the case of the Eocene) to which the early fossils of each group belong. For example, insects make their appearance in fossils of the Devonian Period of the Paleozoic Era almost 400 million years ago. These very early representatives of the insect class were unable to fly.

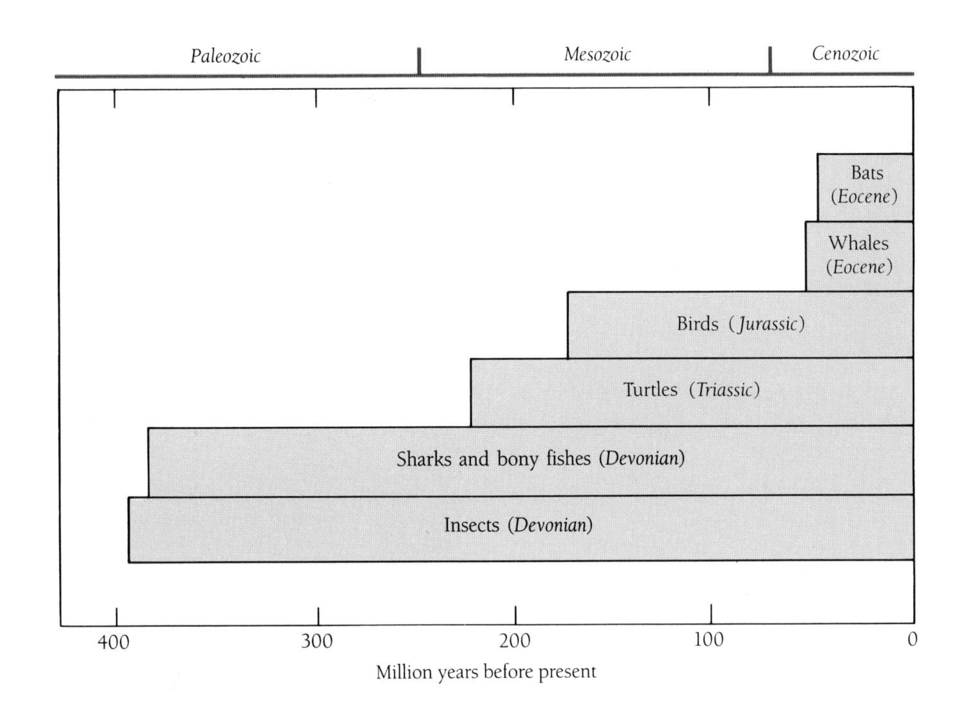

while mountains build, glaciers melt, the earth's magnetic field changes, lakes dry up, continents drift, and weather as well as climate shift drastically. How can migration be perceived within such a kaleidoscope?

Clearly the resulting environmental opportunities and roadblocks challenge animals to adapt both at the short-term behavioral level and at the natural selection level that guides evolution. Consider, for instance, the history of birds, which have proved most adept in exploiting navigation and migration to promote their global success as a group. Most birds evolved between 50 and 199 million years ago, originally from primitive reptilian ancestors. So surviving species must have successfully adapted to the many environmental alterations accompanying global geological changes since then. These geological changes have been remarkably intense. During the time that there have been birds on earth, for instance, the location and relations of the earth's continents have changed. Australia drifted north isolating Antarctica; the old world and the new—once contiguous—moved far apart. In addition the earth underwent considerable cooling 40 million years ago, and the first recent south polar ice appeared about 30 million years ago. Ice also covered the north pole 2.4 million years ago. Only 0.9 million years ago a series of subsequent "ice ages" began, marked by spreading of the polar ice cap southward into middle latitudes such as the British Isles in Western Europe and Long Island in eastern North Amer-

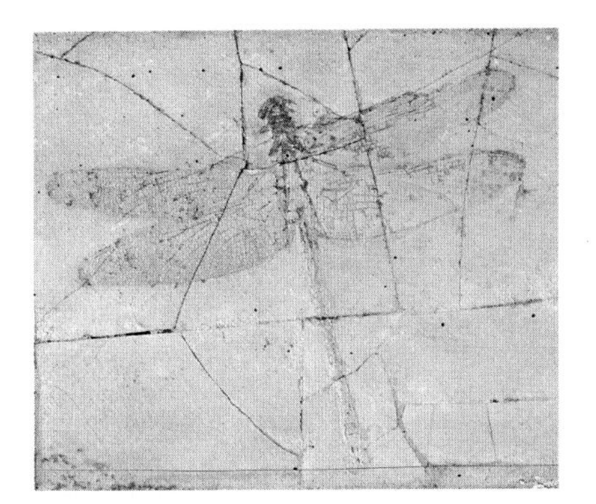

Fossil of a dragonfly of the Jurassic Period. By this time, insects had developed wings that were capable of flight.

Antarctica is covered with a glacier thicker than 3000 meters throughout most of the interior of the continent. The glacier thins along its edges, where the photo was taken. During the last 3 million years, similar huge packs of ice have periodically advanced over northern North America and Europe.

ica. The most recent ice age began about 70,000 years ago and reached cold peaks near 50,000 and 20,000 years ago; then the glaciers retreated. The dates given are quite approximate because actual timing of different glacial and interglacial phases varied in specific parts of the world. The coldest periods were typically punctuated every 10,000 years or so by warm periods lasting perhaps two millenia; the warm interglacials were similarly interrupted by "brief" periods of cold. In some early attempts to explain bird migrations, scientists related them to changing glaciation in the northern hemisphere. Retreat of the ice, due to warming of the earth, was thought to allow high-latitude immigration by originally tropical or temperate types. Or, alternatively, southward advance of the glaciers might have forced the retreat of previously high-latitude species towards the equator. However, existing migratory flyways do not seem to provide much support for such hypotheses.

In any case, current bird migratory patterns are believed to have evolved quite recently, perhaps only 5000 to 15,000 years ago (during the last glacial age's late period of retreat and the beginning of the current interglacial age), which was some time after modern *Homo sapiens* first emerged on earth. This is remarkably rapid as evolution goes and fits the general notion that behavioral evolution can be far quicker than the changes in anatomy reflected in animal classification into species, genera, families, and so on. Indeed there is considerable evidence that the migration of birds is quite flexible and still evolving at an appreciable pace. Bird migratory behavior has clearly changed in response to the destruction of marshlands and rain forests and to natural changes in temperature, rainfall, and ocean currents. What will happen to animal magnetic direction finding when the earth's magnetic field undergoes another of its many reversals that exchange north and south poles? Some evidence suggests that this may occur within the next 2000 years!

Two major theories have been proposed to account for the evolution of massive north-south seasonal migrations by new world and western European land birds. One of these assumes that the ancestral forms were tropical species that discovered the benefits of moving north in summer to acquire new feeding and breeding resources as well as to escape the competition of species-rich low latitudes. In contrast the other theory postulates that migrators originated as temperate birds that learned to move south in winter to avoid hostile climates and seasonal scarcity of food. More recent analysis suggests that neither of these two notions fully explains the known facts. Instead still another alternative suggests that the long-range migratory habit more likely arose in an intermediate staging area near the tropics where an unstable and rather unpredictable environment favored nomadic behavior. A possible scenario could begin with a resident species originally tropical moving into the subtropics. As a result of competition for available resources and climatic changes this then becomes partially migratory. Thereafter it could emerge as a short-range 100-percent migrant and finally as one expanding to a long range.

Continental drift

Adaptive opportunities for migration have surely existed since the origin of animals about 700 million years ago. At various periods in this part of the earth's several-billion-year history, seasonal climates and patchy extremes of environment, including ice caps and glaciation, have been repeatedly present. Strong flyers and swimmers have existed as potential long-range migrators at least since the origin of the earliest groups that have major migrators still living. Insects arose about 400 million years ago; birds appeared about 150 million years ago. Some scientists have tried to associate the evolution of migration with the origin and history of the world's major land masses. A single supercontinent, Pangaea, existed about 250 million years ago. All the present continents can be traced back to the breakup of this huge land mass and the subsequent drifting apart of its components. The problem is that although the behavioral ecology of some animals undoubtedly changed as land masses moved, what early migratory or navigational behaviors might have been is largely conjectural.

One example of a migratory route that may have arisen as continents drifted is that of an extraordinary long-range migration of the marine green turtle *Chelonia mydas*. The species shows a high degree of site fidelity to their breeding places on mid-Atlantic island beaches. These highly restricted areas are several thousand kilometers from this turtle's South American seaweed pasture and even farther from

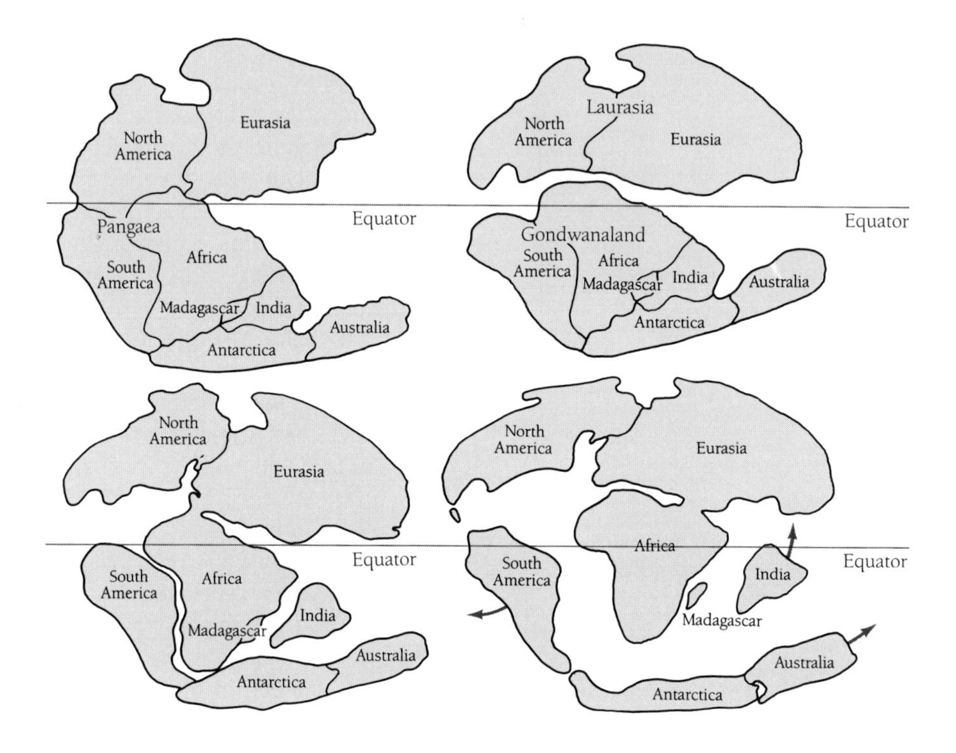

(*Upper left*) Around 250 million years ago most of the earth's land was united into one supercontinent called Pangaea. (*Upper right*) The short-lived supercontinent initially split apart into northern and southern land masses. (*Bottom left to right*) Later the southern continent fragmented, giving birth to the South Atlantic and subsequently widening the ocean.

pelagic regions where they also forage. As this species may take 20 to 30 years to reach sexual maturity, the return of an adult to breed on its natal beach is separated from its departure as a juvenile by two or three decades! Green turtles of certain Brazilian populations swim 2000 kilometers to Ascension Island in the middle of the South Atlantic Ocean for breeding. Zoologists have speculated that this migratory route evolved together with the ancient widening of a sea passage between Africa and South America, which were once fused together. This great rift apparently began 100 million years ago and became nearly as wide as it is now about 1 million years ago. Another Brazilian population of green turtles migrates to less distant mainland coastal breeding places while a closely related species seems to be nonmigratory. Despite the turtles' inherent interest, commercial value, and endangered state, much remains to be learned about their migratory pathways and navigational mechanisms.

Conclusions

Whatever the association of navigation and migration with evolution may finally turn out to be, one broad generalization is certain. Navigational prowess and migration are widespread in the animal kingdom but not ubiquitous—they must have

Brazilian green turtles born on Ascension Island in the center of the South Atlantic spend their lives foraging near the South American coast (blue dots). But the turtles swim back out into the middle of the ocean to breed on their natal island. This unusual migratory route (dashed lines) may have evolved during the rifting of the Atlantic Ocean illustrated on the facing page.

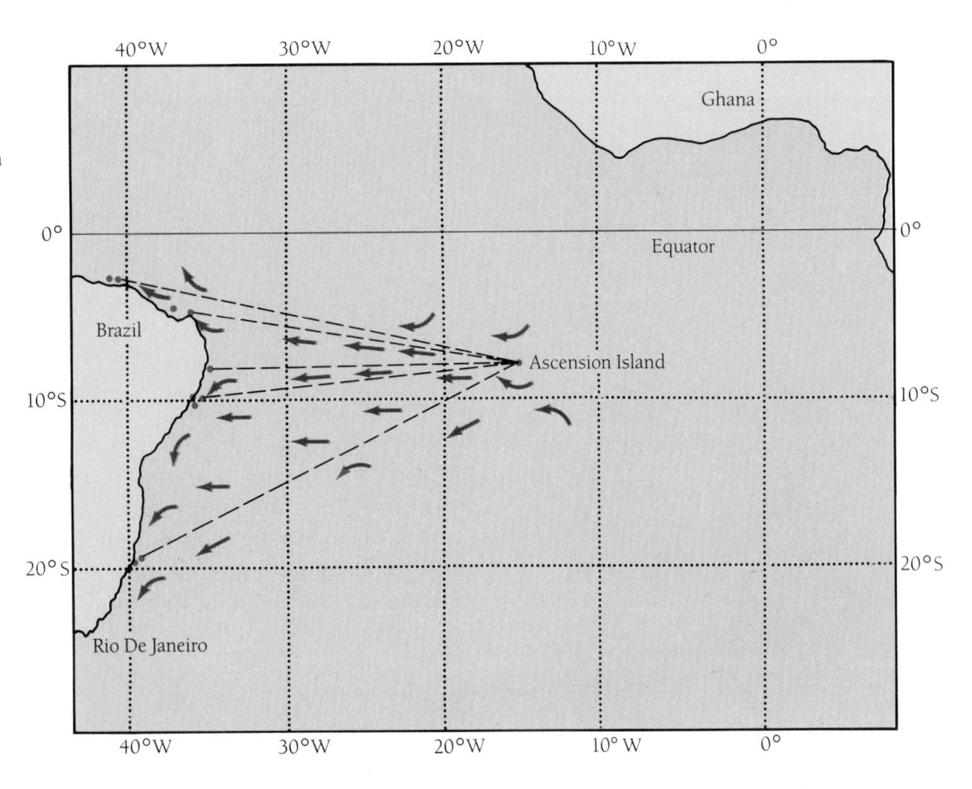

evolved over and over again. Migration may appear to be primarily a response to periodic changes in the environment or it may mainly serve the shifting needs of the life cycle. Yet there seems to be an almost shocking lack of coherence in its distribution. Closely related species of insects, fishes, and birds can be at the extremes of global traveler or stay-at-home. The monarch butterfly of eastern North America seems unique among the huge array of insects in the remarkable extent and specificity of its overwintering migration. Even within species, populations, sexes, age groups, or even among individuals, there may be variation between being migratory or not. Navigational and migrational skills appear to be flexible options readily available to many kinds of animals when they need them. This is in contrast, for example, to something structural, such as the paired limbs or gill arches of vertebrates, that is deeply embedded in the group's basic pattern of development and evolution. Although such anatomical elements were adaptable enough to evolve into fins, wings, or walking legs, there is a stability and conservatism to them quite absent from such behavior as migration.

Green turtle

One may well wonder, for instance, what to make of the homing pigeon. In wild individuals there is no evidence for unusual navigational powers. Yet domesticated and trained birds show astonishing ability to return to their loft after large displacements to previously unknown release points. How they do it cannot yet be satisfactorily explained. Also there is still no general proof that migratory species in nature can reliably perform navigation comparable to that of the pigeon. Does this mean that homing pigeons are quite special in having strongly developed skills which they never use in nature? Or does it mean that highly sophisticated navigation is within the ability of many animals, but this pigeon species for some reason has proved to be the only one to yield certain of its secrets? Stay tuned!

CODA

To end this inquiry into animal navigation we might reasonably ask whether some general conclusions can be drawn from current knowledge. If a simple universal explanation is required, the answer is "no." The solutions to most problems about animal navigation are not simple, and many of them are still to be found. Actually some parts of the story about *how* animals navigate can be described and understood rather well from a functional and biophysical point of view. For instance, certain mechanisms based on familiar senses operating alone, like sight or hearing, can be explained in a reasonably satisfactory way. Even so, their more esoteric aspects, like polarized-light orientation or ultrasonic communication underwater, still have important and conspicuous gaps to be filled. When it comes to altogether less familiar (or perhaps even dubious) sensory functions relating, say, to infrasound or weak magnetic fields, current knowledge is scrappy indeed.

The deficiencies become even more serious if we remember that animals' perceptions of place, direction, and time are ordinarily multimodal with several senses and mechanisms being involved. When needed, every scrap of available information seems to be put to use. The whole topic of how convergent data channels function together is only beginning to be studied effectively. We recognize important principles of "attention" and assigning priorities to various kinds of signals but have little understanding of how they apply to navigation. Certainly, the total sensory, learned, and genetic information available to the animal must be highly "distilled" to allow the best choices for action. Thousands of incoming signals must be evaluated, yet the resulting decisions, like *turn right,* or *swim faster,* or *stop,* are elementary. Clearly mood and motivation may have strong shaping functions in animal behavior. We learned for example, that certain tropical freshwater fish become more alert visually when they smell food in the water. In addition, attention to this chemical signal may range from near 0 to 100 percent depending on the animal's "sleep-wake" state. Despite scattered areas of good understanding, the overall picture remains elusive. New insights would be welcome.

Actually the map needed for animal navigation is probably the least known component and the most difficult to analyze of the basic requisites. Direction finding and timing, by comparison, seem rather straightforward. Yet their usefulness de-

Snow geese over Tule Lake, California, on the West Coast flyway.

pends on having a map. Except when the area involved can be sensed directly, map elements must be either innate or learned, which makes them hard to study. A number of researchers studying homing and migration find this space sense the most stubborn of their problems. Clearly more effort should be devoted to the topic. Indeed one could argue that nearly all aspects of animal navigation both need and deserve similar attention.

One striking aspect of our subject is the contrast between its apparent narrowness as a discipline and the actual breadth of its interest and substance, including the many kinds of knowledge needed to understand it. This paradox no doubt arises from a deep connectedness in nature—its ultimate *unity*. An animal's sun compass, for instance, depends not only on the celestial mechanics of the solar system, but also on the molecular structure of the organism's own DNA (and numberless other things in between!). In a way then, understanding even a small part of the universe requires a deep appreciation of the whole. If we really become aware of this, we know that apparently trivial happenings often lead to profound, far-reaching effects. Much subtle and nonlinear interplay underlies cosmic reality. Evidence for this could come to you when next you see a monarch butterfly fluttering south along a windy Atlantic beach or a persistent salmon flinging its way up a white-water cataract to spawn. As nature's witness, you will sense what I mean if you feel a twinge of recognition and realize you have touched the infinite.

Selected Reading

Introduction

Baker, R. R. *The Evolutionary Ecology of Animal Migration*. New York: Holmes and Meier, 1978.

Gauthreaux, S. A., Jr. *Animal Migration, Orientation, and Navigation*. New York: Academic, 1980.

Hewson, B. A. *A History of the Practice of Navigation,* 2d ed. Glasgow: Brown, Son and Ferguson, 1983.

Maloney, E. S. *Dutton's Navigation and Piloting,* 13th ed. Annapolis, Maryland: Naval Institute Press, 1978.

Wohlschlag, D. E., ed. *Contributions in Marine Science,* vol. 27: September 1985. Port Arkansas, Texas: Marine Science Institute, The University of Texas at Austin, 1985.

Chapter 1 Animal Migration: Flyers

Ackery, P. R. and R. I. Van-Wright. *Milkweed Butterflies*. London: British Museum, 1984.

Danthanarayana, W., ed. *Insect Flight: Dispersal and Migration*. Berlin: Springer-Verlag, 1986.

Hill, J. E. and J. D. Smith. *Bats: A Natural History*. London: British Museum, 1984.

Johnsgard, P. A. *Hummingbirds of North America*. Washington, D. C.: Smithsonian Institute Press, 1983.

Papi, F. and H. G. Wallraff. *Avian Navigation*. Berlin: Springer-Verlag, 1982.

Salomonsen, F. "Migratory movements of the arctic tern (*Sterna paradisaea* Pontoppidan) on the southern ocean." Copenhagen: *Kongelige Danske Videnskabernes Selskab. Biologiske Meddelsen* 24(1), 3–42. (1967).

Williams, T. C. and J. M. Williams. "An oceanic mass migration of land birds." *Scientific American* 239(4), 166–176 (1967).

Chapter 2 Animal Migration: Swimmers

Arnold, G. P. "Movements of fish in relation to water currents." In *Animal Migration* (ed. D. J. Aidley), pp. 55–79. Cambridge: Cambridge University Press, 1981.

McCleave, J. D., G. P. Arnold, J. J. Dodson, and W. H. Neill, eds. *Mechanisms of Migration in Fishes.* New York: Plenum Press, 1984.

Payne, R., ed. *Communication and Behavior of Whales.* Boulder, Colorado: Westview, 1983.

Tesch, F.-W. *The Eel: Biology and Management of Anguillid Eels.* New York: Wiley, 1973.

Würsig, B. "The behavior of baleen whales." *Scientific American* 258(4), 102–107 (1988).

Chapter 3 Human Navigation

Finney, B. R. *Hokule'a: The Way to Tahiti.* New York: Dodd, Mead. 1979.

Gladwin, T. *East is a Big Bird.* Cambridge, Mass.: Harvard University Press, 1970.

Hewsom, J. B. *A History of the Practice of Navigation,* 2d ed. Glasgow: Brown, Son and Ferguson, 1983.

Lewis, D. *We the Navigators.* Honolulu: University of Hawaii Press, 1972.

Lewis, D. *The Voyaging Stars.* Sydney: Collins, 1978.

Menard, H. W. *Islands.* New York: Scientific American Library, 1986.

Stroup, E. D. "Navigating without instruments: the voyage of Hokule'a." *Oceanus* 28:69–75 (1985).

Thomas, S. D. "The sons of Pululap: Navigating without instruments." *Oceanus* 28:52–58 (1985).

Chapter 4 Spatial Orientation and Course Keeping

Schöne, H. *Spatial Orientation.* Princeton, N. J.: Princeton University Press, 1984.

Wehner, R. "Spatial vision in arthropods." In *Handbook of Sensory Physiology,* vol. VII/6C, (ed. H. Autrum), pp. 287–616. Berlin: Springer-Verlag, 1981.

Chapter 5 The Compass and Visual Direction Finding

Emlen, S. T. "The stellar-orientation system of a migratory bird." *Scientific American* 233(2), 102–111 (1975).

von Frisch, K. *The Dance Language and Orientation of Bees.* Cambridge, Mass.: Harvard University Press, 1967.

Lythgoe, J. N. *Ecology of Vision.* Oxford: Clarendon Press, 1979.

Neuweiler, G. "Foraging, echolocation and audition in bats." *Naturwiss.* 71:446–455 (1984).

Waterman, T. H. "Natural polarized light and vision." In *Photoreception and Vision in Invertebrates* (ed. M. A. Ali), pp. 63–114. New York: Plenum, 1984.

Waterman, T. H. "Polarization sensitivity." In *Handbook of Sensory Physiology,* vol. VII/6B, (ed. H Autrum), pp. 281–469 Berlin: Springer-Verlag, 1981.

Wehner, R. and S. Rossel. "The bee's celestial compass—A case study in behavioural neurobiology." In *Fortschritte du Zoologie*. Bd. 31. (ed. B. Holldobler and M. Lindauer), pp. 11–53. Stuttgart: Fischer Verlag, 1985.

Chapter 6 Three More Senses for Direction Finding

Alter, R. A. "Models for echolocation." In *Animal Sonar Systems* (ed. R.-G. Busnel and J. E. Fish), pp. 625–671. New York: Plenum, 1980.

Bleckman, H. In *Progress in Sensory Physiology,* vol 5. (ed. D. Ottoson), pp. 147–166. Berlin: Springer-Verlag, 1985.

Busnel, R.-G. and J. E. Fish. *Animal Sonar Systems*. New York: Plenum, 1980.

Döving, K. B. "Functional properties of the fish olfactory system." In *Progress in Sensory Physiology,* vol. 5. (ed. D. Ottoson), pp. 39–104. Berlin: Springer-Verlag, 1985.

Gaskin, D. E. *The Ecology of Whales and Dolphins*. London: Heinemann, 1982.

Hasler, A. D. *Underwater Guideposts*. Madison, Wis.: The University of Wisconsin Press, 1966.

Payne, T. L., M. C. Birch, and C. E. J. Kennedy, eds. *Mechanisms in Insect Olfaction*. Oxford: Clarendon Press, 1986.

Chapter 7 Electric and Magnetic Direction Finding

Bullock, T. H. and W. Heiligenberg, eds. *Electroreception*. New York: Wiley, 1986.

Kalmijn, A. J. "The detection of electric fields from inanimate and animate sources other than electric organs." In *Handbook of Sensory Physiology,* vol. III/3, (ed. A Fessard), pp. 145–157. Berlin: Springer-Verlag, 1974.

Kirschvink, J. L., D. S. Jones, and B. J. MacFadden, eds. *Magnetite Biomineralisation and Magnetoreception in Organisms,* New York: Plenum, 1985.

Maret, G., J. Kiepenheuer, and N. Boccara, eds. *Biophysical Effects of Steady Magnetic Fields*. Berlin: Springer-Verlag, 1986.

Semm, P., D. Nohr, C. Demaine, and W. Wiltschko. "Neural basis of the magnetic compass: Interactions of visual, magnetic and vestibular inputs in the pigeon's brain." *Journal of Comparative Physiology A* 155, 283–288 (1984).

Wiltschko, W. and R. Wiltschko. "Magnetic orientation in birds." in *Current Ornithology,* vol. 5, (ed. R. F. Johnson), pp. 67–121. New York: Plenum, 1988.

Chapter 8 Sense of Space: The Map

Archer, J. and L. Birke, eds. *Exploration in Animals and Humans*. Van Nostrand Reinhold (UK), 1983.

Barnes, C. A. and B. L. McNaughton, "Spatial information: how and where is it stored?" In *Memory Systema of the Brain. Animal and Human Cognitive Processes* (ed. N. M. Weinberger, J. L. McGaugh, and G. Lynch), pp. 49–61. New York: Guilford, 1985.

Cartwright, B. A. and T. S. Collett. "Landmark learning in bees. Experiments and models." *Journal of Comparative Physiology A* 151(4), 521–544 (1983).

Coss, R. G. "The function of dendritic spines: A review of theoretical issues." *Behavioral and Neurological Biology* 44, 151–185 (1985).

Fawcett, J. W. and D. D. M. O'Leary. "The role of electrical activity in the formation of topographic maps in the nervous system. In *Trends in Neuroscience* 8(5), 201–206 (1985).

Gould, J. L. and C. G. Gould, *The Honey Bee*. New York: Scientific American Library, 1988.

Konishi, M. "Centrally synthesized maps of sensory space." *Trends in Neuroscience* 9(4), 163–168 (1986).

Schmid-Hempel, P. "Foraging characteristics of the desert ant *Cataglyphis*." In *Experientia Supplementum*, vol 54: *Behavior in Social Insects*, pp. 43–61. Basel: Birkhaüser Verlag, 1987.

Sherry, D. E. "Food storage by birds and mammals." *Advances in the Study of Behavior* 15, 153–188 (1984).

Wehner, R. and M. V. Srinivasan. "Searching behaviour of desert ants, genus *Cataglyphis* (Formicidae, Hymenoptera)." *Journal of Comparative Physiology A* 142, 315–338 (1981).

Chapter 9 Sense of Time: Biological Clocks and Calendars

Aschoff, J., ed. *Handbook of Sensory Phsyiology*, vol. 4: *Biological Rhythms*. New York: Plenum, 1981.

Binkley, S. "A time keeping enzyme in the pineal gland." *Scientific American* 240(4), 66–71 (1979).

Connor, J. A. "Neural pacemakers and rhythmicity." *Annual Review of Physiology* 47, 17–29 (1985).

Gwinner, E. "Circannual rhythms in the control of avian migration." *Advances in the Study of Behavior* 16, 191–228 (1985).

Konopka, R. J. "Genetics of biological rhythms in *Drosophila*." *Annual Advances in Genetics* 21, 227–234 (1987).

Olson, L. M. and J. W. Jacklet. "The circadian pacemaker in the *Aplysia* eye sends axons throughout the central nervous system." *Journal of Neuroscience* 5(12), 3214–3227 (1985).

Saunders, D. S. *Insect Clocks*, 2d ed. Oxford: Oxford University Press, 1982.

Tauber, M. J., C. A. Tauber, and S. Masaki. *Seasonal Adaptation of Insects*. New York: Oxford University Press, 1986.

Willis, J. M. and W. G. Pearcy. "Vertical distribution and migration of fishes of the lower mesopelagic zone off Oregon." *Marine Biology* 70, 87–98 (1982).

Chapter 10 Why Animals Migrate

Barnard, C. J. *Animal Behaviour. Ecology and Evolution.* New York: Wiley, 1983.

Dingle, H. "Ecology and evolution." In *Animal Migration: Orientation and Navigation* (ed. S. A. Gauthreaux, Jr.), pp. 1–101. New York: Academic, 1980.

Dingle, H. "Evolution and genetics of insect migration." In *Insect Flight: Dispersal and Migration* (ed. W. Danthanarayana), pp. 11–26. Berlin: Springer-Verlag, 1986.

Donaldson, L. R., and T. Joyner. "The salmonid fishes as a natural livestock." *Scientific American* 240(1), 50–58 (1983).

Gross, M. R., R. M. Coleman, and R. M. McDowall. "Aquatic productivity and the evolution of diadromous fish migration." *Science* 239, 1291–1293 (1988).

Scapini, F. "Inheritance of direction finding in sandhoppers." In *Orientation in Space* (ed. G. Beugnon), pp. 111–119. Toulouse: Privat, I.E.C., 1986.

Simpson, G. G. *Fossils and the History of Life.* New York: Scientific American Library, 1983.

Stanley, S. M. *Extinction.* New York: Scientific American Library, 1987.

Stowe, M. K., J. H. Tomlinson, and R. R. Heath. "Chemical mimicry: Bolas spiders emit components of moth prey species sex pheromone." Science 236: 964–967 (1987).

Swingland, I. R. and P. J. Greenwood. *The Ecology of Animal Movement.* Oxford: Clarendon Press, 1983.

Sources of Illustrations

Frontispiece
Frans Lanting

Introduction opener
Frans Lanting

Page 3
Peter Roberts

Page 4
Tupper Ansel Blake

Page 5
(*bottom*) David Sailors

Page 6
Philip Rosenberg

Page 8
Edward Ross

Page 9
Franz J. Camenzind/Planet Earth Pictures

Page 10
Edward Ross

Page 11
Tupper Ansel Blake

Chapter 1 opener
Daniel Maffia

Page 16
Adapted from F. Salomonsen, *Kongelige Danske Videnskabernes Selskab Biologiske Meddelsen* 24 (1967).

Page 17
(*left*) Frans Lanting
(*right*) Adapted from J. C. Welty, The Life of Birds, 3d ed., Saunders, Philadelphia, 1982

Page 18
Adapted from R. T. Orr, *Animal Migration*, Macmillan, New York, 1970

Page 19
Adapted from L. Löfgren, *Ocean Birds*, Knopf, New York, 1984

Page 21
(*top*) Adapted from J. G. Philips et al. (eds.), *Physiological Strategies in Avian Biology*, Blackie & Sons, Glasgow, 1985
(*bottom*) Julian Hector/Planet Earth Pictures

Page 22
Frans Lanting

Page 24
Adapted from A. Keast and E. Morton (eds.), *Migrant Birds in the Neotropics: Ecology, Behavior, Distribution, and Conservation,* Smithsonian, Washington, D.C., 1980

Page 25
Mike Tracey/Survival Anglia

Page 27
Adapted from J. C. Welty, *The Life of Birds*

Page 28
Adapted from J. Dorst, *The Migration of Birds*, London, 1962

Page 30
Merlin D. Tuttle/Bat Conservation International

Page 32
G. Tortoli/Food and Agricultural Organization

Page 33
Adapted from F. Urquhart
Canadian Entomologist 109 (1977)

Page 34
Peter Menzel

Chapter 2 opener
Science (239) 1988; photograph by Marj Trim, courtesy of the Department of Fisheries, Canada and Douglas & McIntyre Ltd.

Page 38
Rick Frehsee, courtesy of William Herrnkind

Page 39
(*top*) Adapted from J. N. Lythgoe, *Handbook of Sensory Physiology* (ed. H. J. A. Dartnall), Springer-Verlag, Berlin, 1972

Page 39
Flip Nicklin

Page 40
Adapted from G. P. Arnold and P. H. Cook, *Mechanisms of Migration in Fishes* (ed. J. D. McCleave, G. P. Arnold, J. J. Dodson, and W. H. Neill), Plenum, New York, 1984

Page 41
Adapted from J. Meinke, *The Undersea* (ed. N. C. Fleming), Macmillan, New York, 1977

Page 43
(*top*) Flip Nicklin
(*bottom*) Adapted from L. R. Rivas, *Physiological Ecology of Tuna* (eds. G. D. Sharp and A. E. Dizon), Academic, New York, 1977

Page 46
Heather Angel

Page 47
Heather Angel

Page 48
Michel Therien, Department of Fisheries and Oceans, Canada

Page 49
Adapted from C. Groot and T. P. Quinn, *Fishery Bulletin* 85 (1967)

Page 50
Flip Nicklin

Page 51
Fançois Gohmer

Page 52
Flip Nicklin

Page 53
Adapted from M. L. Jones, S. Swartz, and S. Leatherhead (eds.), *The Gray Whale*, Academic, New York, 1984

Page 55
Tupper Ansel Blake

Chapter 3 opener
Daniel Maffia

Page 58
Adapted from S. D. Thomas, "The Sons of Pululap: Navigating without instruments," *Oceanus* 28 (1985)

Page 59
Bernice P. Bishop Museum, Honolulu

Page 60
S. D. Thomas, *The Last Navigator*, Henry Holt & Company, New York, 1987

Page 61
Adapted from B. R. Finney, *The Way to Tahiti*, Dodd, Mead, New York, 1979

Page 62
Adapted from T. Gladwin, *East Is a Big Bird*, Harvard University Press, Cambridge, Mass., 1970

Page 64
Douglas Faulkner

Page 65
Adapted from D. Lewis, *The Voyaging Stars*, Collins, Sydney, 1978

Page 66
(*top*) Adapted from A. F. Aveni, *Science* 213 (1981)
(*bottom left*) U. S. Naval Observatory
(*bottom right*) University of Hawaii Institute for Astronomy

Page 67
Adapted from T. Gladwin, *East Is a Big Bird*

Page 69
Adapted from A. F. Aveni, *Science* 213 (1981)

Chapter 4 opener
Agence Nature/NHPA

Page 75
(*top left*) Flip Nicklin
(*top right*) Carl Roessler

Page 76
Adapted from H. Schöne, *Journal of Comparative Physiology* 107 (1976)

Page 77
Adapted from R. Eckert and D. Randall, *Animal Physiology: Mechanisms and Adaptations*, 2d ed., W. H. Freeman and Company, New York, 1983

Page 78
Adapted from M. J. Wells, *Journal of Experimental Biology* 37 (1960)

Page 79
Adapted from E. von Holst, *Society for Experimental Biology Symposium* 4 (1950)

Page 82
Adapted from A. Flock, *Cold Spring Harbor Symposium* (1965)

Page 84
(*left*) Adapted from H. Mittelstaedt, *Zeitschrift für vergleichende Physiologie* 32 (1950)
(*right*) Stephen Dalton/NHPA

Page 85
Adapted from F. S. J. Hollick, *Philosophical Transactions of the Royal Society* London B 230 (1940) and J. W. S. Pringle, *Philosophical Transactions of the Royal Society* London B 233 (1948)

Page 86
Friedrich Barth

Page 90
Adapted from H. Heran and M. Lindauer, *Zeitschrift für vergleichende Physiologie* 47 1963

Chapter 5 opener
Stephen Dalton/NHPA

Page 97
(*left*) Adapted from Wagner, 1896
(*right*) Adapted from M. F. Land and T. S. Collet, *Journal of Comparative Physiology* 89 (1974)

Page 98
Rüdiger Wehner

Page 100
Adapted from A. Hasler, *Underwater Guideposts*, University of Wisconsin Press, Madison, Wis., 1966

Page 102
Adapted from F. G. Barth, *Insects and Flowers*, Princeton University Press, Princeton, N.J., 1985

Page 103
Kenneth Lorenzen

Page 107
Adapted from E. S. Maloney, *Dutton's Navigation and Piloting*, Naval Institute Press, Annapolis, Md., 1978

Page 109
(top) Jonathan Blair/Woodfin Camp & Associates
(bottom) Adapted from A. C. Fisher, *Mysteries of Bird Migration*, 1979

Page 111
L. Pardi

Page 114
(top) Edward Ross

Page 115
(left) K. Kirschfeld, *Naturwissenschaftliche Rundschau* 37/9 (1984)

Page 116
(left) John Shaw
(top middle) T. H. Waterman and K. W. Horch, *Science* 154 (1966)
(bottom middle) T. H. Waterman, *Handbook of Sensory Physiology,* vol. VII/6B (ed. H. Autrum), Springer-Verlag, Berlin, 1981
(right) E. Eguchi and T. H. Waterman, *Functional Organization of the Compound Eye* (ed. C. G. Bernhard), Pergamon, Oxford,

Page 117
Adapted from K. Daumer, *Zeitschrift für vergleichende Physiologie* 38 (1956)

Page 119
Heather Angel

Page 120
Adapted from T. H. Waterman, *Science* 111 (1950)

Page 121
P. J. Bryant, University of California, Irvine/BPS

Chapter 6 opener
Daniel Maffia

Page 124
Dwight Kuhn

Page 125
(left) Heather Angel
(right) Adapted from Blaxter et al., *Journal of the Marine Biological Association*, U.K., 63 (1983)

Page 126
Adapted from H. Margenau et al., *Physics*, Saunders, Philadephia, 1949

Page 129
(top) Stephen Dalton/NHPA
(bottom) Adapted from D. Griffin, *Scientific American* 199 (1958)

Page 130
Merlin D. Tuttle/Bat Conservation International

Page 131
Edward Ross

Page 132
(left) Adapted from Pilleri, *Endavour* NS 7
(right) Adapted from R. Baker, *The Evolutionary Ecology of Animal Migration*, Holmes & Meier, New York, 1978

Page 133
Adapted from W. W. L. Au and C. W. Turl, *Journal of Acoustic Society of America* 73

Page 134
(top and middle) Adapted from F. G. Barth, *Insects and Flowers*
(bottom) P. J. Bryant, U. of California, Irvine/BPS

Page 135
(left) Adapted from Hoelldobler et al., *Behavior, Ecology, and Sociobiology* 4 (1978)

Page 136
(left) G. I. Bernard/Oxford Scientific Films
(right) R. A. Steinbrecht

Page 137
Department of Fisheries and Oceans, Canada

Page 142
(top) Ken Lucas/BPS
(bottom) Adapted from E. Newman and P. Hartline, *Scientific American* 246 (1982)

Page 143
Adapted from E. Newman and P. Hartline, *Scientific American* 246 (1982)

Chapter 7 opener
Daniel Maffia

Page 146
Dennis Willows

Page 147
Adapted from H. W. Lissmann *Scientific American* (208) 1963

Page 148
Jean-Paul Ferrero/AUSCAPE

Page 149
Adapted from Szabo, *Handbook of Sensory Physiology,* vol. III/3, Springer-Verlag, Berlin, 1974

Page 150
Adapted from W. A. Raschi, *Journal of Morphology* 189 (1986)

Page 151
Robert J. Erwin/NHPA

Page 153
Adapted from R. Eckert and D. Randall, *Animal Physiology: Mechanisms and Adaptations*, 2d ed.

Page 154
Adapted from B. Dutton, *Navigation and Nautical Astronomy*, 10th ed., U.S. Naval Institute, Annapolis, Md. 1951

Page 155
Adapted from K. Schmidt-Koenig, *Avian Orientation and Navigation*, Academic, New York, 1979

Page 156
Adapted from J. L. Kirschvink et al. (eds.), *Magnetite Biomineralization and Magnetoreception in Organisms*, Plenum, New York, 1985

Page 158
(top) Adapted from M. Lindauer and H. Martin, "Magnetic effect on dancing bees," Symposium NASA SP 262
(bottom) Adaptation from C. Mead, *Bird Migration*, Facts on File, New York, 1983

Page 159
Wolfgang Wagner/Ardea, London

Page 162
Richard Chesher/Planet Earth Pictures

Page 164
(*top*) Adapted from A. Kalmijn, *Handbook of Sensory Physiology*, vol. III/3, Springer-Verlag, Berlin, 1974
(*bottom*) D. Maratea, courtesy of Nancy Blakemore

Page 166
Lang Elliott

Chapter 8 opener
George Lepp/Comstock

Page 170
Adapted from T. Seeley, *Scientific American*, 1983

Page 171
Robert Tyrrell

Page 172
Adapted from P. Knoppien and J. Reddingius, 1985

Page 173
Tupper Ansel Blake

Page 175
Russ Charif, courtesy of Charles Walcott

Page 176
Adapted from R. T. Orr, *Animal Migration*

Page 178
Adapted from K. Schmidt-Koenig, "Bird orientation and navigation"

Page 179
National Maritime Museum, Greenwich, England

Page 180
C. Plath

Page 184
Adapted from J. T. Bonner, *Evolution of Culture in Animals*, Princeton University Press, Princeton, N.J., 1980

Page 186
Tupper Ansel Blake

Page 187
Adapted from R. Wehner and M. V. Srinivasan, *Journal of Comparative Physiology A* 142 (1981)

Page 188
Edward Ross

Chapter 9 opener
J. P. Bryant, University of California, Irvine/BPS

Page 192
Adapted from J. Aschoff (ed.), *Handbook of Behavioral Neurobiology*, vol. 4, Plenum, New York, 1981

Page 193
Hugh Yorkston

Page 165
(*left*) John Shaw
(*right*) Adapted from G. Fleissner, *Naturwissenschaften* 73 (1986)

Page 197
(*left*) Peter David/Planet Earth Pictures
(*right*) Peter Parks/Oxford Scientific Films

Page 198
Adapted from Palmer, *Scientific American* (1975)

Page 199
Jeff Foott/Survival Anglia

Page 200
Adapted from D. S. Heintzelman, *The Migrations of Hawks*, Indiana University Press, Bloomington, Ind., 1986

Page 201
Adapted from J. Aschoff (ed.), *Handbook of Behavioral Neurobiology*, vol. 4

Page 202
National Maritime Museum, Greenwich, England

Page 203
Charles Arneson

Page 205
Edward Ross

Chapter 10 opener
Frans Lanting

Page 208
Stephen Dalton/NHPA

Page 209
(*top*) Merlin D. Tuttle/Bat Conservation International
(*bottom*) Alan Root/Survival Anglia

Page 210
Jonathan Scott/Planet Earth Pictures

Page 212
Patricia Biesiot and Judith McDowell Capuzzo

Page 217
J. A. Bailey/Ardea, London

Page 219
(*bottom*) Field Museum of Natural History

Page 220
Bora Merdsoy/Planet Earth Pictures

Page 222
L. E. M. De Boer, B. Van Wissen, "Notogaea" 8 (1977)

Page 223
(*top*) Archie Carr et al., *Science*
(*bottom*) Flip Nicklin

Page 225
Tupper Ansel Blake

Index

Other books in the Scientific American Library Series

POWERS OF TEN
by Philip and Phylis Morrison and
the Office of Charles and Ray Eames

HUMAN DIVERSITY
by Richard Lewontin

THE DISCOVERY OF SUBATOMIC PARTICLES
by Steven Weinberg

THE SCIENCE OF MUSICAL SOUND
by John R. Pierce

FOSSILS AND THE HISTORY OF LIFE
by George Gaylord Simpson

THE SOLAR SYSTEM
by Roman Smoluchowski

ON SIZE AND LIFE
by Thomas A. McMahon and John Tyler Bonner

PERCEPTION
by Irvin Rock

CONSTRUCTING THE UNIVERSE
by David Layzer

THE SECOND LAW
by P. W. Atkins

THE LIVING CELL, VOLUMES I AND II
by Christian de Duve

MATHEMATICS AND OPTIMAL FORM
by Stefan Hildebrandt and Anthony Tromba

FIRE
by John W. Lyons

SUN AND EARTH
by Herbert Friedman

EINSTEIN'S LEGACY
by Julian Schwinger

ISLANDS
by H. William Menard

DRUGS AND THE BRAIN
by Solomon H. Snyder

THE TIMING OF BIOLOGICAL CLOCKS
by Arthur T. Winfree

EXTINCTION
by Steven M. Stanley

MOLECULES
by P. W. Atkins

EYE, BRAIN, AND VISION
by David H. Hubel

THE SCIENCE OF STRUCTURES AND MATERIALS
by J. E. Gordon

SAND
by Raymond Siever

THE HONEY BEE
by James L. Gould and Carol Grant Gould